PLASMONICS: FUNDAMENTALS AND APPLICATIONS

PLASMONICS: FUNDAMENTALS AND APPLICATIONS

STEFAN A. MAIER
Centre for Photonics and Photonic Materials
Department of Physics, University of Bath, UK

Stefan A. Maier
Centre for Photonics & Photonic Materials
Department of Physics
University of Bath
Bath BA2 7AY
United Kingdom

Plasmonics: Fundamentals and Applications

Library of Congress Control Number: 2006931007

ISBN 0-387-33150-6 e-ISBN 0-387-37825-1
ISBN 978-0387-33150-8 e-ISBN 978-0387-37825-1

Printed on acid-free paper.

9 8 7 6 5 4 3 2 1

springer.com

For Harry Atwater, with
thanks for the great time.

Contents

List of Figures

Foreword

It was the autumn of 1982 and my final year undergraduate project was on surface plasmons. I had no idea that this topic would still have me fascinated almost a quarter of a century later, let alone have become a life-time career. Time really does fly. The invitation to write a foreword to this book with the instruction that it include a historical perspective set me thinking of my own first encounter with surface plasmons. My project supervisor was Roy Sambles - little did I realise how lucky I was. Without knowing it I became hooked on physics; not just studying it but doing it - I was off. The field of surface plasmons has changed enormously in the intervening years; indeed, in its new guise as plasmonics, interest has soared and many more people have joined the field.

But for those new to the topic, where to begin? A good book can act as a guide and companion - it can make all the difference. When I started in 1982 the newest book was a monster, a compilation called "Electromagnetic Surface Waves", edited by Alan Boardman. Together with Kevin Welford, I had joined Roy Sambles to do a PhD - as beginners we found this book a daunting yet valuable resource – we plundered it, before long the pages became dog-eared and the covers fell off. I left things plasmonic in 1986, not to rejoin until 1992. In the meantime Hans Raether published "Surface Plasmons". With his wonderful combination of simplicity and insight, especially in the introductory sections, a classic emerged. Now almost twenty years later it is still very much in use but, inevitably, it has become increasingly out of date as the field continues to rapidly expand. Whilst several specialist volumes have emerged, we have been acutely aware of the need for a more up-to-date introduction and overview of the field at a glance. Now we have it - thank you Stefan.

But what is plasmonics? "You just have Maxwell's equations, some material properties and some boundary conditions, all classical stuff - what's new about that?" Well, would you have predicted that just by imposing appropriate structure on a metal one could make a synthetic material that would turn Snell's law on its head? Or that you could squeeze light into places less that one hundredth of a wavelength in size? No new fundamental particles, no new cosmology - but surprises, adventure, the quest to understand - yes, we have all of those, and more.

It seems that four elements underlie research in plasmonics today. The first is the ready availability of state-of-the-art fabrication methods, particularly for implementing nanostructure. Second, there are a wealth of high-sensitivity optical characterisation techniques, which one can buy pretty much off-the-shelf. Third, the rapid advance in computing power and speed have allowed us to implement powerful numerical modelling tools on little more than a laptop computer. The fact that many researchers can gain access to these things enables the expansion of the field of plasmonics, but what has motivated that expansion?

The cynic might argue fashion. However, the fourth element, the one missing from the list above, is the wide range of potential applications - solar cells, high-resolution microscopy, drug design and many more. Applications are indeed strong motivators, but I think there is more to it than that. I know I am biased, but for me and I suspect many others it's the adventure, the role of the imagination, the wish to be the one to find something new, to explain the unexplained - in short its science, simple as that. Perhaps amazingly there are still many topics in which one can do all of these things without the need to observe gravity waves, build particle accelerators, or even work out how the brain that loves to do such things works. Plasmonics is one of those small-scale topics where good people can do interesting things with modest resources, that too is one of the lures.

Roughly speaking the field is a hundred years old. Around the turn of the last century the same four elements as described above applied - albeit in a different way. The relevant state-of-the-art fabrication was that of ruled diffraction gratings, optical characterisation was provided by the same gratings - to give spectroscopy. Computation was based on, among others, Rayleigh's work on diffraction and Zenneck's and Sommerfeld's work on surface waves - all analytical, but still valuable today. There was in addition an improved understanding of metals, particularly from Drude's treatment. So what was missing? Perhaps most importantly these different activities were not really recognised as having a commonality in the concept of surface plasmons. Now we are in a very different situation, one in which the relevant underlying science is much better understood - but where, as we continue to see, there are still many surprises.

Looking back it seems clear that the 1998 paper in Nature by Thomas Ebbesen and colleagues on the extraordinary transmission of light through metallic hole-arrays triggered many to enter the field. With an avalanche of developments in spectral ranges from the microwave, through THz, IR and visible, and into the UV the need for an entry point has become more acute. Well, here it is.

It can't possibly be comprehensive, but Stefan Maier's addition gives an up-to-date introduction and a great overview of the present situation. Who knows what new concepts might emerge and where the important applications will be? Maybe none of us know yet, that's the beauty - it could be you.

Bill Barnes,
School of Physics, University of Exeter,
June 2006

Preface

Plasmonics forms a major part of the fascinating field of *nanophotonics*, which explores how electromagnetic fields can be confined over dimensions on the order of or smaller than the wavelength. It is based on interaction processes between electromagnetic radiation and conduction electrons at metallic interfaces or in small metallic nanostructures, leading to an enhanced optical near field of sub-wavelength dimension.

Research in this area demonstrates how a distinct and often unexpected behavior can occur (even with for modern optical studies seemingly uninteresting materials such as metals!) if discontinuities or sub-wavelength structure is imposed. Another beauty of this field is that it is firmly grounded in classical physics, so that a solid background knowledge in electromagnetism at undergraduate level is sufficient to understand main aspects of the topic.

However, history has shown that despite the fact that the two main ingredients of plasmonics - *surface plasmon polaritons* and *localized surface plasmons* - have been clearly described as early as 1900, it is often far from trivial to appreciate the interlinked nature of many of the phenomena and applications of this field. This is compounded by the fact that throughout the 20th century, surface plasmon polaritons have been rediscovered in a variety of different contexts.

The mathematical description of these surface waves was established around the turn of the 20th century in the context of radio waves propagating along the surface of a conductor of finite conductivity [Sommerfeld, 1899, Zenneck, 1907]. In the visible domain, the observation of *anomalous* intensity drops in spectra produced when visible light reflects at metallic gratings [Wood, 1902] was not connected with the earlier theoretical work until mid-century [Fano, 1941]. Around this time, loss phenomena associated with interactions taking place at metallic surfaces were also recorded via the diffraction of electron beams at thin metallic foils [Ritchie, 1957], which was in the 1960s then linked with the original work on diffraction gratings in the optical domain [Ritchie et al., 1968]. By that time, the excitation of Sommerfeld's surface waves with visible light using prism coupling had been achieved [Kretschmann and Raether, 1968], and a unified description of all these phenomena in the form of surface plasmon polaritons was established.

From then on, research in this field was so firmly grounded in the visible region of the spectrum, that several rediscoveries in the microwave and the terahertz domain took place at the turn of the 21st century, closing the circle with the original work from 100 years earlier. The history of localized surface plasmons in metal nanostructures is less turbulent, with the application of metallic nanoparticles for the staining of glass dating back to Roman times. Here, the clear mathematical foundation was also established around 1900 [Mie, 1908].

It is with this rich history of the field in mind that this book is written. It is aimed both at students with a basic undergraduate knowledge in electromagnetism or applied optics that want to start exploring the field, and at researchers as a hopefully valuable desk reference. Naturally, this necessitates an extensive reference section. Throughout the book, the original studies described and cited were selected either because they provided to the author's knowledge the first description of a particular effect or application, or due to their didactic suitability at the point in question. In many cases, it is clear that also different articles could have been chosen, and in some sections of the book only a small number of studies taken from a pool of qualitatively similar work had to be selected.

The first part of this text should provide a solid introduction into the field, starting with an elementary description of classic electromagnetism, with particular focus on the description of conductive materials. Subsequent chapters describe both surface plasmon polaritons and localized plasmons in the visible domain, and electromagnetic surface modes at lower frequencies. In the second part, this knowledge is applied to a number of different applications, such as plasmon waveguides, aperture arrays for enhanced light transmission, and various geometries for surface-enhanced sensing. The book closes with a short description of metallic metamaterials.

I hope this text will serve its purpose and provide a useful tool for both current and future participants in this area, and will strengthen a feeling of community between the different sub-fields. Comments and suggestions are very much appreciated.

STEFAN MAIER

Acknowledgments

I wish to thank my colleague Tim Birks for all his efforts in proof-reading an early draft of this book and his helpful criticism, and David Bird for his encouragement to undertake this project. Thanks also to my student Charles de Nobriga for working through a more advanced version of this text, and of course to my wife Mag for all the lovely distractions from writing ...

PART I

FUNDAMENTALS OF PLASMONICS

Introduction

Research in plasmonics is currently taking place at a breathtaking pace, and we can expect that many more will join the game in the near future. But for the newcomer, where to start? Before diving into particular sub-fields, fundamental or application-driven, a solid basis for an understanding of the more specialized literature is clearly desirable. This part of the text aims to help in building such a core knowledge. The first chapter sets the groundwork by describing the optical properties of metals, starting with Maxwell's equations and the derivation of the dielectric function of the free electron gas. The following three chapters introduce surface plasmon polaritons both at single interfaces and in multilayer structures, and describe experimental techniques for their excitation and observation. Chapter 5 adds the second important ingredient of the game, localized surface plasmons in metallic nanostructures. The first part of the book closes by describing electromagnetic modes at low frequencies, where surface plasmon polaritons based on metals become highly delocalized, and surface structuring must be employed to create more confined modes.

Chapter 1

ELECTROMAGNETICS OF METALS

While the optical properties of metals are discussed in most textbooks on condensed matter physics, for convenience this chapter summarizes the most important facts and phenomena that form the basis for a study of surface plasmon polaritons. Starting with a cursory review of Maxwell's equations, we describe the electromagnetic response both of idealized and real metals over a wide frequency range, and introduce the fundamental excitation of the conduction electron sea in bulk metals: *volume plasmons*. The chapter closes with a discussion of the electromagnetic energy density in dispersive media.

1.1 Maxwell's Equations and Electromagnetic Wave Propagation

The interaction of metals with electromagnetic fields can be firmly understood in a classical framework based on Maxwell's equations. Even metallic nanostructures down to sizes on the order of a few nanometres can be described without a need to resort to quantum mechanics, since the high density of free carriers results in minute spacings of the electron energy levels compared to thermal excitations of energy $k_{\mathrm{B}}T$ at room temperature. The optics of metals described in this book thus falls within the realms of the classical theory. However, this does not prevent a rich and often unexpected variety of optical phenomena from occurring, due to the strong dependence of the optical properties on frequency.

As is well known from everyday experience, for frequencies up to the visible part of the spectrum metals are highly reflective and do not allow electromagnetic waves to propagate through them. Metals are thus traditionally employed as cladding layers for the construction of waveguides and resonators for electromagnetic radiation at microwave and far-infrared frequencies. In this low-frequency regime, the *perfect* or *good conductor approximation* of infinite or fixed finite conductivity is valid for most purposes, since only a negligible fraction of the impinging electromagnetic waves penetrates into the metal. At higher frequencies towards the near-infrared and visible part of the spectrum, field penetration increases significantly, leading to increased dissipation, and prohibiting a simple size scaling of photonic devices that work well at low frequencies to this regime. Finally, at ultraviolet frequencies, metals acquire dielectric character and allow the propagation of electromagnetic waves, albeit with varying degrees of attenuation, depending on the details of the electronic band structure. Alkali metals such as sodium have an almost free-electron-like response and thus exhibit an *ultraviolet transparency*. For noble metals such as gold or silver on the other hand, transitions between electronic bands lead to strong absorption in this regime.

These dispersive properties can be described via a complex dielectric function $\varepsilon(\omega)$, which provides the basis of all phenomena discussed in this text. The underlying physics behind this strong frequency dependence of the optical response is a change in the phase of the induced currents with respect to the driving field for frequencies approaching the reciprocal of the characteristic *electron relaxation time* τ of the metal, as will be discussed in section 1.2.

Before presenting an elementary description of the optical properties of metals, we recall the basic equations governing the electromagnetic response, the *macroscopic Maxwell equations*. The advantage of this *phenomenological approach* is that details of the fundamental interactions between charged particles inside media and electromagnetic fields need not be taken into account, since the rapidly varying microscopic fields are averaged over distances much larger than the underlying microstructure. Specifics about the transition from a microscopic to a macroscopic description of the electromagnetic response of continuous media can be found in most textbooks on electromagnetics such as [Jackson, 1999].

We thus take as a starting point Maxwell's equations of macroscopic electromagnetism in the following form:

$$\nabla \cdot \mathbf{D} = \rho_{\text{ext}} \tag{1.1a}$$

$$\nabla \cdot \mathbf{B} = 0 \tag{1.1b}$$

$$\nabla \times \mathbf{E} = -\frac{\partial \mathbf{B}}{\partial t} \tag{1.1c}$$

$$\nabla \times \mathbf{H} = \mathbf{J}_{\text{ext}} + \frac{\partial \mathbf{D}}{\partial t}. \tag{1.1d}$$

These equations link the four macroscopic fields **D** (the dielectric displacement), **E** (the electric field), **H** (the magnetic field), and **B** (the magnetic induction or magnetic flux density) with the external charge and current densities ρ_{ext} and $\mathbf{J}_{\text{ext}}$. Note that we do not follow the usual procedure of presenting the macroscopic equations via dividing the total charge and current densities ρ_{tot} and $\mathbf{J}_{\text{tot}}$ into free and bound sets, which is an arbitrary division [Illinskii and Keldysh, 1994] and can (especially in the case of metallic interfaces) confuse the application of the boundary condition for the dielectric displacement. Instead, we distinguish between external $(\rho_{\text{ext}}, \mathbf{J}_{\text{ext}})$ and internal $(\rho, \mathbf{J})$ charge and current densities, so that in total $\rho_{\text{tot}} = \rho_{\text{ext}} + \rho$ and $\mathbf{J}_{\text{tot}} = \mathbf{J}_{\text{ext}} + \mathbf{J}$. The external set drives the system, while the internal set responds to the external stimuli [Marder, 2000].

The four macroscopic fields are further linked via the polarization **P** and magnetization **M** by

$$\mathbf{D} = \varepsilon_0 \mathbf{E} + \mathbf{P} \tag{1.2a}$$

$$\mathbf{H} = \frac{1}{\mu_0}\mathbf{B} - \mathbf{M}, \tag{1.2b}$$

where ε_0 and μ_0 are the electric permittivity[1] and magnetic permeability[2] of vacuum, respectively. Since we will in this text only treat nonmagnetic media, we need not consider a magnetic response represented by **M**, but can limit our description to electric polarization effects. **P** describes the electric dipole moment per unit volume inside the material, caused by the alignment of microscopic dipoles with the electric field. It is related to the internal charge density via $\nabla \cdot \mathbf{P} = -\rho$. Charge conservation $(\nabla \cdot \mathbf{J} = -\partial \rho / \partial t)$ further requires that the internal charge and current densities are linked via

$$\mathbf{J} = \frac{\partial \mathbf{P}}{\partial t}. \tag{1.3}$$

The great advantage of this approach is that the macroscopic electric field includes all polarization effects: In other words, both the external and the induced fields are absorbed into it. This can be shown via inserting (1.2a) into (1.1a), leading to

$$\nabla \cdot \mathbf{E} = \frac{\rho_{\text{tot}}}{\varepsilon_0}. \tag{1.4}$$

In the following, we will limit ourselves to linear, isotropic and nonmagnetic media. One can define the constitutive relations

$$\mathbf{D} = \varepsilon_0 \varepsilon \mathbf{E} \tag{1.5a}$$

$$\mathbf{B} = \mu_0 \mu \mathbf{H}. \tag{1.5b}$$

[1] $\varepsilon_0 \approx 8.854 \times 10^{-12}$ F/m
[2] $\mu_0 \approx 1.257 \times 10^{-6}$ H/m

ε is called the dielectric constant or relative permittivity and $\mu = 1$ the relative permeability of the nonmagnetic medium. The linear relationship (1.5a) between **D** and **E** is often also implicitly defined using the dielectric susceptibility χ (particularly in quantum mechanical treatments of the optical response [Boyd, 2003]), which describes the linear relationship between **P** and **E** via

$$\mathbf{P} = \varepsilon_0 \chi \mathbf{E}. \tag{1.6}$$

Inserting (1.2a) and (1.6) into (1.5a) yields $\varepsilon = 1 + \chi$.

The last important constitutive linear relationship we need to mention is that between the internal current density **J** and the electric field **E**, defined via the conductivity σ by

$$\mathbf{J} = \sigma \mathbf{E}. \tag{1.7}$$

We will now show that there is an intimate relationship between ε and σ, and that electromagnetic phenomena with metals can in fact be described using either quantity. Historically, at low frequencies (and in fact in many theoretical considerations) preference is given to the conductivity, while experimentalists usually express observations at optical frequencies in terms of the dielectric constant. However, before embarking on this we have to point out that the statements (1.5a) and (1.7) are only correct for linear media that do not exhibit temporal or spatial dispersion. Since the optical response of metals clearly depends on frequency (possibly also on wave vector), we have to take account of the non-locality in time and space by generalizing the linear relationships to

$$\mathbf{D}(\mathbf{r}, t) = \varepsilon_0 \int dt' \mathbf{dr}' \varepsilon(\mathbf{r} - \mathbf{r}', t - t') \mathbf{E}(\mathbf{r}', t') \tag{1.8a}$$

$$\mathbf{J}(\mathbf{r}, t) = \int dt' \mathbf{dr}' \sigma(\mathbf{r} - \mathbf{r}', t - t') \mathbf{E}(\mathbf{r}', t'). \tag{1.8b}$$

$\varepsilon_0 \varepsilon$ and σ therefore describe the impulse response of the respective linear relationship. Note that we have implicitly assumed that all length scales are significantly larger than the lattice spacing of the material, ensuring homogeneity, i.e. the impulse response functions do not depend on absolute spatial and temporal coordinates, but only their differences. For a local response, the functional form of the impulse response functions is that of a δ-function, and (1.5a) and (1.7) are recovered.

Equations (1.8) simplify significantly by taking the Fourier transform with respect to $\int dt \mathbf{dr} e^{i(\mathbf{K} \cdot \mathbf{r} - \omega t)}$, turning the convolutions into multiplications. We are thus decomposing the fields into individual plane-wave components of wave vector **K** and angular frequency ω. This leads to the constitutive relations in the Fourier domain

$$\mathbf{D}(\mathbf{K},\omega) = \varepsilon_0 \varepsilon(\mathbf{K},\omega)\mathbf{E}(\mathbf{K},\omega) \tag{1.9a}$$

$$\mathbf{J}(\mathbf{K},\omega) = \sigma(\mathbf{K},\omega)\mathbf{E}(\mathbf{K},\omega). \tag{1.9b}$$

Using equations (1.2a), (1.3) and (1.9) and recognizing that in the Fourier domain $\partial/\partial t \to -i\omega$, we finally arrive at the fundamental relationship between the relative permittivity (from now on called the *dielectric function*) and the conductivity

$$\varepsilon(\mathbf{K},\omega) = 1 + \frac{i\sigma(\mathbf{K},\omega)}{\varepsilon_0 \omega}. \tag{1.10}$$

In the interaction of light with metals, the general form of the dielectric response $\varepsilon(\omega, \mathbf{K})$ can be simplified to the limit of a *spatially local* response via $\varepsilon(\mathbf{K} = \mathbf{0}, \omega) = \varepsilon(\omega)$. The simplification is valid as long as the wavelength λ in the material is significantly longer than all characteristic dimensions such as the size of the unit cell or the mean free path of the electrons. This is in general still fulfilled at ultraviolet frequencies[3].

Equation (1.10) reflects a certain arbitrariness in the separation of charges into bound and free sets, which is entirely due to convention. At low frequencies, ε is usually used for the description of the response of bound charges to a driving field, leading to an electric polarization, while σ describes the contribution of free charges to the current flow. At optical frequencies however, the distinction between bound and free charges is blurred. For example, for highly-doped semiconductors, the response of the bound valence electrons could be lumped into a static dielectric constant $\delta\varepsilon$, and the response of the conduction electrons into σ', leading to a dielectric function $\varepsilon(\omega) = \delta\varepsilon + \frac{i\sigma'(\omega)}{\varepsilon_0\omega}$. A simple redefinition $\delta\varepsilon \to 1$ and $\sigma' \to \sigma' + \frac{\varepsilon_0\omega}{i}\delta\varepsilon$ will then result in the general form (1.10) [Ashcroft and Mermin, 1976].

In general, $\varepsilon(\omega) = \varepsilon_1(\omega) + i\varepsilon_2(\omega)$ and $\sigma(\omega) = \sigma_1(\omega) + i\sigma_2(\omega)$ are complex-valued functions of angular frequency ω, linked via (1.10). At optical frequencies, ε can be experimentally determined for example via reflectivity studies and the determination of the complex refractive index $\tilde{n}(\omega) = n(\omega) + i\kappa(\omega)$ of the medium, defined as $\tilde{n} = \sqrt{\varepsilon}$. Explicitly, this yields

[3]However, spatial dispersion effects can lead to small corrections for surface plasmons polaritons in metallic nanostructures significantly smaller than the electron mean free path, which can arise for example at the tip of metallic cones (see chapter 7).

$$\varepsilon_1 = n^2 - \kappa^2 \tag{1.11a}$$

$$\varepsilon_2 = 2n\kappa \tag{1.11b}$$

$$n^2 = \frac{\varepsilon_1}{2} + \frac{1}{2}\sqrt{\varepsilon_1^2 + \varepsilon_2^2} \tag{1.11c}$$

$$\kappa = \frac{\varepsilon_2}{2n}. \tag{1.11d}$$

κ is called the extinction coefficient and determines the optical absorption of electromagnetic waves propagating through the medium. It is linked to the absorption coefficient α of Beer's law (describing the exponential attenuation of the intensity of a beam propagating through the medium via $I(x) = I_0 e^{-\alpha x}$) by the relation

$$\alpha(\omega) = \frac{2\kappa(\omega)\omega}{c}. \tag{1.12}$$

Therefore, the imaginary part ε_2 of the dielectric function determines the amount of absorption inside the medium. For $|\varepsilon_1| \gg |\varepsilon_2|$, the real part n of the refractive index, quantifying the lowering of the phase velocity of the propagating waves due to polarization of the material, is mainly determined by ε_1. Examination of (1.10) thus reveals that the real part of σ determines the amount of absorption, while the imaginary part contributes to ε_1 and therefore to the amount of polarization.

We close this section by examining traveling-wave solutions of Maxwell's equations in the absence of external stimuli. Combining the curl equations (1.1c), (1.1d) leads to the *wave equation*

$$\nabla \times \nabla \times \mathbf{E} = -\mu_0 \frac{\partial^2 \mathbf{D}}{\partial t^2} \tag{1.13a}$$

$$\mathbf{K}(\mathbf{K} \cdot \mathbf{E}) - K^2\mathbf{E} = -\varepsilon(\mathbf{K}, \omega)\frac{\omega^2}{c^2}\mathbf{E}, \tag{1.13b}$$

in the time and Fourier domains, respectively. $c = \frac{1}{\sqrt{\varepsilon_0 \mu_0}}$ is the speed of light in vacuum. Two cases need to be distinguished, depending on the polarization direction of the electric field vector. For transverse waves, $\mathbf{K} \cdot \mathbf{E} = 0$, yielding the generic dispersion relation

$$K^2 = \varepsilon(\mathbf{K}, \omega)\frac{\omega^2}{c^2}. \tag{1.14}$$

For longitudinal waves, (1.13b) implies that

$$\varepsilon(\mathbf{K}, \omega) = 0, \tag{1.15}$$

signifying that longitudinal collective oscillations can only occur at frequencies corresponding to zeros of $\varepsilon(\omega)$. We will return to this point in the discussion of volume plasmons in section 1.3.

1.2 The Dielectric Function of the Free Electron Gas

Over a wide frequency range, the optical properties of metals can be explained by a *plasma model*, where a gas of free electrons of number density n moves against a fixed background of positive ion cores. For alkali metals, this range extends up to the ultraviolet, while for noble metals interband transitions occur at visible frequencies, limiting the validity of this approach. In the plasma model, details of the lattice potential and electron-electron interactions are not taken into account. Instead, one simply assumes that some aspects of the band structure are incorporated into the effective optical mass m of each electron. The electrons oscillate in response to the applied electromagnetic field, and their motion is damped via collisions occurring with a characteristic collision frequency $\gamma = 1/\tau$. τ is known as the relaxation time of the free electron gas, which is typically on the order of 10^{-14} s at room temperature, corresponding to $\gamma = 100$ THz.

One can write a simple equation of motion for an electron of the plasma sea subjected to an external electric field $\mathbf{E}$:

$$m\ddot{\mathbf{x}} + m\gamma\dot{\mathbf{x}} = -e\mathbf{E} \tag{1.16}$$

If we assume a harmonic time dependence $\mathbf{E}(t) = \mathbf{E}_0 \mathrm{e}^{-i\omega t}$ of the driving field, a particular solution of this equation describing the oscillation of the electron is $\mathbf{x}(t) = \mathbf{x}_0 \mathrm{e}^{-i\omega t}$. The complex amplitude $\mathbf{x}_0$ incorporates any phase shifts between driving field and response via

$$\mathbf{x}(t) = \frac{e}{m(\omega^2 + i\gamma\omega)}\mathbf{E}(t). \tag{1.17}$$

The displaced electrons contribute to the macroscopic polarization $\mathbf{P} = -ne\mathbf{x}$, explicitly given by

$$\mathbf{P} = -\frac{ne^2}{m(\omega^2 + i\gamma\omega)}\mathbf{E}. \tag{1.18}$$

Inserting this expression for $\mathbf{P}$ into equation (1.2a) yields

$$\mathbf{D} = \varepsilon_0(1 - \frac{\omega_\mathrm{p}^2}{\omega^2 + i\gamma\omega})\mathbf{E}, \tag{1.19}$$

where $\omega_\mathrm{p}^2 = \frac{ne^2}{\varepsilon_0 m}$ is the *plasma frequency* of the free electron gas. Therefore we arrive at the desired result, the dielectric function of the free electron gas:

$$\varepsilon(\omega) = 1 - \frac{\omega_p^2}{\omega^2 + i\gamma\omega}. \tag{1.20}$$

The real and imaginary components of this complex dielectric function $\varepsilon(\omega) = \varepsilon_1(\omega) + i\varepsilon_2(\omega)$ are given by

$$\varepsilon_1(\omega) = 1 - \frac{\omega_p^2\tau^2}{1 + \omega^2\tau^2} \tag{1.21a}$$

$$\varepsilon_2(\omega) = \frac{\omega_p^2\tau}{\omega(1 + \omega^2\tau^2)}, \tag{1.21b}$$

where we have used $\gamma = 1/\tau$. It is insightful to study (1.20) for a variety of different frequency regimes with respect to the collision frequency γ. We will limit ourselves here to frequencies $\omega < \omega_p$, where metals retain their metallic character. For large frequencies close to ω_p, the product $\omega\tau \gg 1$, leading to negligible damping. Here, $\varepsilon(\omega)$ is predominantly real, and

$$\varepsilon(\omega) = 1 - \frac{\omega_p^2}{\omega^2} \tag{1.22}$$

can be taken as the dielectric function of the undamped free electron plasma. Note that the behavior of noble metals in this frequency region is completely altered by interband transitions, leading to an increase in ε_2. The examples of gold and silver will be discussed below and in section 1.4.

We consider next the regime of very low frequencies, where $\omega \ll \tau^{-1}$. Hence, $\varepsilon_2 \gg \varepsilon_1$, and the real and the imaginary part of the complex refractive index are of comparable magnitude with

$$n \approx \kappa = \sqrt{\frac{\varepsilon_2}{2}} = \sqrt{\frac{\tau\omega_p^2}{2\omega}}. \tag{1.23}$$

In this region, metals are mainly absorbing, with an absorption coefficient of

$$\alpha = \left(\frac{2\omega_p^2\tau\omega}{c^2}\right)^{1/2}. \tag{1.24}$$

By introducing the dc-conductivity σ_0, this expression can be recast using $\sigma_0 = \frac{ne^2\tau}{m} = \omega_p^2\tau\varepsilon_0$ to

$$\alpha = \sqrt{2\sigma_0\omega\mu_0}. \tag{1.25}$$

The application of Beer's law of absorption implies that for low frequencies the fields fall off inside the metal as $e^{-z/\delta}$, where δ is the skin depth

$$\delta = \frac{2}{\alpha} = \frac{c}{\kappa\omega} = \sqrt{\frac{2}{\sigma_0 \omega \mu_0}}. \tag{1.26}$$

A more rigorous discussion of the low-frequency behavior based on the Boltzmann transport equation [Marder, 2000] shows that this description is indeed valid as long as the mean free path of the electrons $l = v_F \tau \ll \delta$, where v_F is the Fermi velocity. At room temperature, for typical metals $l \approx 10$ nm and $\delta \approx 100$ nm, thus justifying the free-electron model. At low temperatures however, the mean free path can increase by many orders of magnitude, leading to changes in the penetration depth. This phenomenon is known as the anomalous skin effect.

If we use σ instead of ε for the description of the dielectric response of metals, we recognize that in the absorbing regime it is predominantly real, and the free charge velocity responds in phase with the driving field, as can be seen by integrating (1.17). At DC, relaxation effects of free charges are therefore conveniently described via the real DC-conductivity σ_0, whereas the response of bound charges is put into a dielectric constant ε_B, as discussed above in the examination of the interlinked nature between ε and σ.

At higher frequencies ($1 \leq \omega\tau \leq \omega_p\tau$), the complex refractive index is predominantly imaginary (leading to a reflection coefficient $R \approx 1$ [Jackson, 1999]), and σ acquires more and more complex character, blurring the boundary between free and bound charges. In terms of the optical response, $\sigma(\omega)$ enters expressions only in the combination (1.10) [Ashcroft and Mermin, 1976], due to the arbitrariness of the division between free and bound sets discussed above.

Whereas our description up to this point has assumed an ideal free-electron metal, we will now briefly compare the model with an example of a real metal important in the field of plasmonics (an extended discussion can be found in section 1.4). In the free-electron model, $\varepsilon \to 1$ at $\omega \gg \omega_p$. For the noble metals (e.g. Au, Ag, Cu), an extension to this model is needed in the region $\omega > \omega_p$ (where the response is dominated by free s electrons), since the filled d band close to the Fermi surface causes a highly polarized environment. This residual polarization due to the positive background of the ion cores can be described by adding the term $\mathbf{P}_\infty = \varepsilon_0(\varepsilon_\infty - 1)\mathbf{E}$ to (1.2a), where $\mathbf{P}$ now represents solely the polarization (1.18) due to free electrons. This effect is therefore described by a dielectric constant ε_∞ (usually $1 \leq \varepsilon_\infty \leq 10$), and we can write

$$\varepsilon(\omega) = \varepsilon_\infty - \frac{\omega_p^2}{\omega^2 + i\gamma\omega}. \tag{1.27}$$

The validity limits of the free-electron description (1.27) are illustrated for the case of gold in Fig. 1.1. It shows the real and imaginary components ε_1 and ε_2 for a dielectric function of this type, fitted to the experimentally determined dielectric function of gold [Johnson and Christy, 1972]. Clearly, at visible

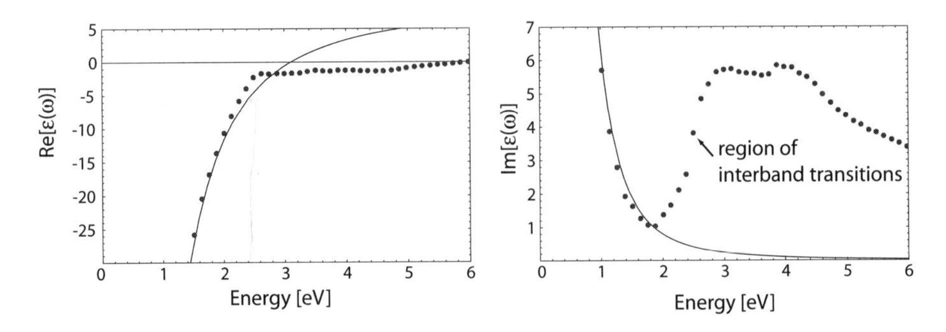

Figure 1.1. Dielectric function $\varepsilon(\omega)$ (1.27) of the free electron gas (solid line) fitted to the literature values of the dielectric data for gold [Johnson and Christy, 1972] (dots). Interband transitions limit the validity of this model at visible and higher frequencies.

frequencies the applicability of the free-electron model breaks down due to the occurrence of interband transitions, leading to an increase in ε_2. This will be discussed in more detail in section 1.4. The components of the complex refractive index corresponding to the fits presented in Fig. 1.1 are shown in Fig. 1.2.

It is instructive to link the dielectric function of the free electron plasma (1.20) to the classical Drude model [Drude, 1900] for the AC conductivity $\sigma(\omega)$ of metals. This can be achieved by recognizing that equation (1.16) can be rewritten as

$$\dot{\mathbf{p}} = -\frac{\mathbf{p}}{\tau} - e\mathbf{E}, \tag{1.28}$$

where $\mathbf{p} = m\dot{\mathbf{x}}$ is the momentum of an individual free electron. Via the same arguments presented above, we arrive at the following expression for the AC conductivity $\sigma = \frac{nep}{m}$,

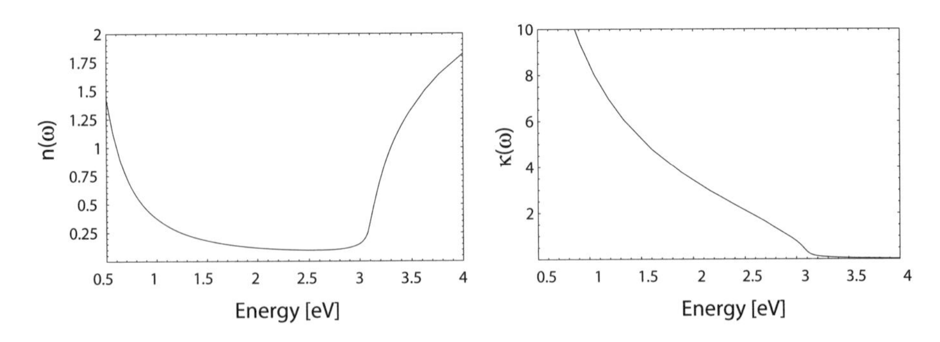

Figure 1.2. Complex refractive index corresponding to the free-electron dielectric function in Fig. 1.1.

$$\sigma(\omega) = \frac{\sigma_0}{1 - i\omega\tau}. \tag{1.29}$$

By comparing equation (1.20) and (1.29), we get

$$\varepsilon(\omega) = 1 + \frac{i\sigma(\omega)}{\varepsilon_0 \omega}, \tag{1.30}$$

recovering the previous, general result of equation 1.10. The dielectric function of the free electron gas (1.20) is thus also known as the Drude model of the optical response of metals.

1.3 The Dispersion of the Free Electron Gas and Volume Plasmons

We now turn to a description of the thus-far omitted transparency regime $\omega > \omega_p$ of the free electron gas model. Using equation (1.22) in (1.14), the dispersion relation of traveling waves evaluates to

$$\omega^2 = \omega_p^2 + K^2 c^2. \tag{1.31}$$

This relation is plotted for a generic free electron metal in Fig. 1.3. As can be seen, for $\omega < \omega_p$ the propagation of transverse electromagnetic waves is forbidden inside the metal plasma. For $\omega > \omega_p$ however, the plasma supports transverse waves propagating with a group velocity $v_g = d\omega/dK < c$.

The significance of the plasma frequency ω_p can be further elucidated by recognizing that in the small damping limit, $\varepsilon(\omega_p) = 0$ (for $\mathbf{K} = 0$). This excitation must therefore correspond to a collective longitudinal mode as shown in the discussion leading to (1.15). In this case, $\mathbf{D} = 0 = \varepsilon_0 \mathbf{E} + \mathbf{P}$. We see that

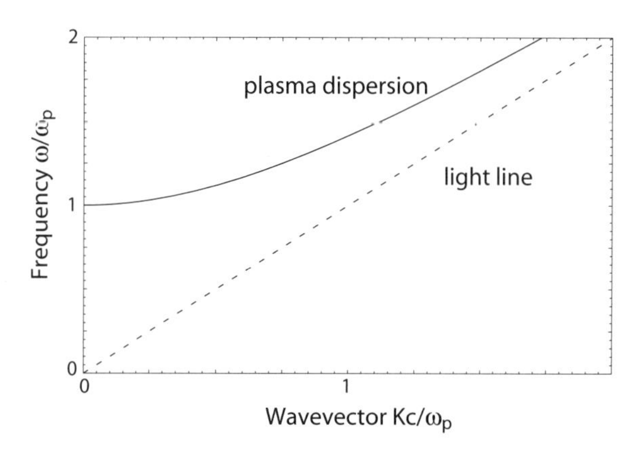

Figure 1.3. The dispersion relation of the free electron gas. Electromagnetic wave propagation is only allowed for $\omega > \omega_p$.

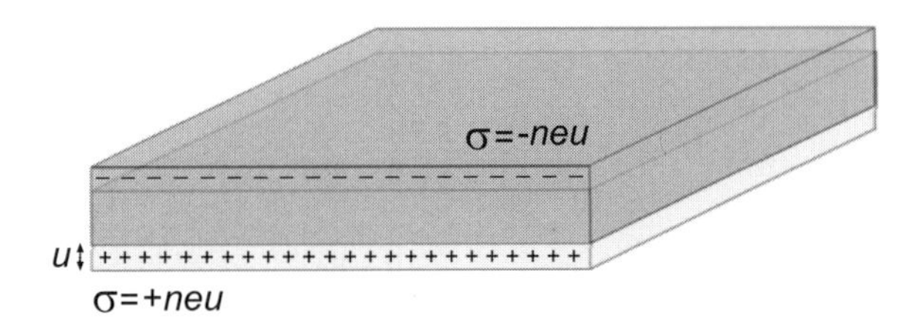

Figure 1.4. Longitudinal collective oscillations of the conduction electrons of a metal: Volume plasmons

at the plasma frequency the electric field is a pure depolarization field, with $\mathbf{E} = \frac{-\mathbf{P}}{\varepsilon_0}$.

The physical significance of the excitation at ω_p can be understood by considering the collective longitudinal oscillation of the conduction electron gas versus the fixed positive background of the ion cores in a plasma slab. Schematically indicated in Fig. 1.4, a collective displacement of the electron cloud by a distance u leads to a surface charge density $\sigma = \pm neu$ at the slab boundaries. This establishes a homogeneous electric field $\mathbf{E} = \frac{neu}{\varepsilon_0}$ inside the slab. Thus, the displaced electrons experience a restoring force, and their movement can be described by the equation of motion $nm\ddot{u} = -ne\mathbf{E}$. Inserting the expression for the electric field, this leads to

$$nm\ddot{u} = -\frac{n^2e^2u}{\varepsilon_0} \tag{1.32a}$$

$$\ddot{u} + \omega_p^2 u = 0. \tag{1.32b}$$

The plasma frequency ω_p can thus be recognized as the natural frequency of a free oscillation of the electron sea. Note that our derivation has assumed that all electrons move in phase, thus ω_p corresponds to the oscillation frequency in the long-wavelength limit where $\mathbf{K} = 0$. The quanta of these charge oscillations are called plasmons (or *volume* plasmons, to distinguish them from *surface* and *localized* plasmons, which will be discussed in the remainder of this text). Due to the longitudinal nature of the excitation, volume plasmons do not couple to transverse electromagnetic waves, and can only be excited by particle impact. Another consequence of this is that their decay occurs only via energy transfer to single electrons, a process known as Landau damping.

Experimentally, the plasma frequency of metals typically is determined via electron loss spectroscopy experiments, where electrons are passed through thin metallic foils. For most metals, the plasma frequency is in the ultraviolet regime: ω_p is on the order of $5 - 15$ eV, depending on details of the band structure [Kittel, 1996]. As an aside, we want to note that such longitudinal oscillations can also be excited in dielectrics, in which case the valence electrons oscillate collectively with respect to the ion cores.

In addition to the in-phase oscillation at ω_p, there exists a whole class of longitudinal oscillations at higher frequencies with finite wavevectors, for which (1.15) is fulfilled. The derivation of the dispersion relation of volume plasmons is beyond the scope of this treatment and can be found in many textbooks on condensed matter physics [Marder, 2000, Kittel, 1996]. Up to quadratic order in $\mathbf{K}$,

$$\omega^2 = \omega_p^2 + \frac{6E_F K^2}{5m}, \tag{1.33}$$

where E_F is the Fermi energy. Practically, the dispersion can be measured using inelastic scattering experiments such as electron energy loss spectroscopy (EELS).

1.4 Real Metals and Interband Transitions

We have already on several occasions stated that the dielectric function (1.20) of the Drude model adequately describes the optical response of metals only for photon energies below the threshold of transitions between electronic bands. For some of the noble metals, interband effects already start to occur for energies in excess of 1 eV (corresponding to a wavelength $\lambda \approx 1\ \mu$m). As examples, Figs. 1.1 and 1.5 show the real and the imaginary parts $\varepsilon_1(\omega)$, $\varepsilon_2(\omega)$ of the dielectric function for gold and silver [Johnson and Christy, 1972] and Drude model fits to the data. Clearly, this model is not adequate for describing either ε_1 or ε_2 at high frequencies, and in the case of gold, its validity breaks down already at the boundary between the near-infrared and the visible.

We limit this comparison between the Drude model and the dielectric response of real metals to the cases of gold and silver, the most important metals for plasmonic studies in the visible and near-infrared. Above their respective

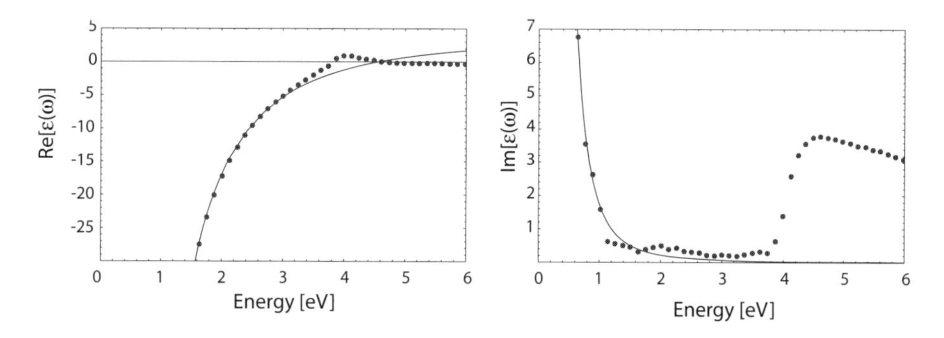

Figure 1.5. The real and imaginary part of $\varepsilon(\omega)$ for silver determined by Johnson and Christy [Johnson and Christy, 1972] (dots) and a Drude model fit to the data.

band edge thresholds, photons are very efficient in inducing interband transitions, where electrons from the filled band below the Fermi surface are excited to higher bands. Theoretically, these can be described using the same approach used for direct band transitions in semiconductors [Ashcroft and Mermin, 1976, Marder, 2000], and we will not embark on a more detailed discussion. The main consequence of these processes concerning surface plasmon polaritons is an increased damping and competition between the two excitations at visible frequencies.

For practical purposes, a big advantage of the Drude model is that it can easily be incorporated into time-domain based numerical solvers for Maxwell's equations, such as the finite-difference time-domain (FDTD) scheme [Kashiwa and Fukai, 1990], via the direct calculation of the induced currents using (1.16). Its inadequacy in describing the optical properties of gold and silver at visible frequencies can be overcome by replacing (1.16) by

$$m\ddot{\mathbf{x}} + m\gamma\dot{\mathbf{x}} + m\omega_0^2\mathbf{x} = -e\mathbf{E}. \tag{1.34}$$

Interband transitions are thus described using the classical picture of a bound electron with resonance frequency ω_0, and (1.34) can then be used to calculate the resulting polarization. We note that a number of equations of this form might have to be solved (each resulting in a separate contribution to the total polarization) in order to model $\varepsilon(\omega)$ for noble metals accurately. Each of these equations leads to a Lorentz-oscillator term of the form $\frac{A_i}{\omega_i^2 - \omega^2 - i\gamma_i\omega}$ added to the free-electron result (1.20) [Vial et al., 2005].

1.5 The Energy of the Electromagnetic Field in Metals

We finish this chapter by taking a brief look at the energy of the electromagnetic field inside metals, or more generally inside dispersive media. Since the amount of field localization is often quantified in terms of the electromagnetic energy distribution, a careful consideration of the effects of dispersion is necessary. For a linear medium with no dispersion or losses (i.e. (1.5) holds), the total energy density of the electromagnetic field can be written as [Jackson, 1999]

$$u = \frac{1}{2}(\mathbf{E} \cdot \mathbf{D} + \mathbf{B} \cdot \mathbf{H}). \tag{1.35}$$

This expression enters together with the Poynting vector of energy flow $\mathbf{S} = \mathbf{E} \times \mathbf{H}$ into the conservation law

$$\frac{\partial u}{\partial t} + \nabla \cdot \mathbf{S} = -\mathbf{J} \cdot \mathbf{E}, \tag{1.36}$$

relating changes in electromagnetic energy density to energy flow and absorption inside the material.

In the following, we will concentrate on the contribution u_E of the electric field $\mathbf{E}$ to the total electromagnetic energy density. In metals, ε is complex and frequency-dependent due to dispersion, and (1.35) does not apply. For a field consisting of monochromatic components, Landau and Lifshitz have shown that the conservation law (1.36) can be held up if u_E is replaced by an *effective* electric energy density u_eff, defined as

$$u_\mathrm{eff} = \frac{1}{2}\mathrm{Re}\left[\frac{d(\omega\varepsilon)}{d\omega}\right]_{\omega_0} \langle\mathbf{E}(\mathbf{r},t)\cdot\mathbf{E}(\mathbf{r},t)\rangle, \qquad (1.37)$$

where $\langle\mathbf{E}(\mathbf{r},t)\cdot\mathbf{E}(\mathbf{r},t)\rangle$ signifies field-averaging over one optical cycle, and ω_0 is the frequency of interest. This expression is valid if $\mathbf{E}$ is only appreciable in a narrow frequency range around ω_0, and the fields are slowly-varying compared to a timescale $1/\omega_0$. Furthermore, it is assumed that $|\varepsilon_2| \ll |\varepsilon_1|$, so that absorption is small. We note that additional care must be taken with the correct calculation of absorption on the right side of (1.36), where $\mathbf{J}\cdot\mathbf{E}$ should be replaced by $\omega_0\mathrm{Im}\left[\varepsilon(\omega_0)\right]\langle\mathbf{E}(\mathbf{r},t)\cdot\mathbf{E}(\mathbf{r},t)\rangle$ if the dielectric response of the metal is completely described via $\varepsilon(\omega)$ [Jackson, 1999], in line with the discussion surrounding (1.10).

The requirement of low absorption limits (1.37) to visible and near-infrared frequencies, but not to lower frequencies or the regime of interband effects where $|\varepsilon_2| > |\varepsilon_1|$. However, the electric field energy can also be determined by taking the electric polarization explicitly into account, in the form described by (1.16) [Loudon, 1970, Ruppin, 2002]. The obtained expression for the electric field energy of a material described by a free-electron-type dielectric function $\varepsilon = \varepsilon_1 + i\varepsilon_2$ of the form (1.20) is

$$u_\mathrm{eff} = \frac{\varepsilon_0}{4}\left(\varepsilon_1 + \frac{2\omega\varepsilon_2}{\gamma}\right)|\mathbf{E}|^2, \qquad (1.38)$$

where an additional factor $1/2$ is included due to an implicit assumption of harmonic time dependence of the oscillating fields. For negligible ε_2, it can be shown that (1.38) reduces as expected to (1.37) for time-harmonic fields. We will use (1.38) in chapter 2 when discussing the amount of energy localization in fields localized at metallic surfaces.

Chapter 2

SURFACE PLASMON POLARITONS AT METAL / INSULATOR INTERFACES

Surface plasmon polaritons are electromagnetic excitations propagating at the interface between a dielectric and a conductor, evanescently confined in the perpendicular direction. These electromagnetic surface waves arise via the coupling of the electromagnetic fields to oscillations of the conductor's electron plasma. Taking the wave equation as a starting point, this chapter describes the fundamentals of surface plasmon polaritons both at single, flat interfaces and in metal/dielectric multilayer structures. The surface excitations are characterized in terms of their dispersion and spatial profile, together with a detailed discussion of the quantification of field confinement. Applications of surface plasmon polaritons in waveguiding will be deferred to chapter 7.

2.1 The Wave Equation

In order to investigate the physical properties of surface plasmon polaritons (SPPs), we have to apply Maxwell's equations (1.1) to the flat interface between a conductor and a dielectric. To present this discussion most clearly, it is advantageous to cast the equations first in a general form applicable to the guiding of electromagnetic waves, the *wave equation*.

As we have seen in chapter 1, in the absence of external charge and current densities, the curl equations (1.1c, 1.1d) can be combined to yield

$$\nabla \times \nabla \times \mathbf{E} = -\mu_0 \frac{\partial^2 \mathbf{D}}{\partial t^2}. \tag{2.1}$$

Using the identities $\nabla \times \nabla \times \mathbf{E} \equiv \nabla(\nabla \cdot \mathbf{E}) - \nabla^2 \mathbf{E}$ as well as $\nabla \cdot (\varepsilon \mathbf{E}) \equiv \mathbf{E} \cdot \nabla \varepsilon + \varepsilon \nabla \cdot \mathbf{E}$, and remembering that due to the absence of external stimuli $\nabla \cdot \mathbf{D} = 0$, (2.1) can be rewritten as

$$\nabla\left(-\frac{1}{\varepsilon}\mathbf{E}\cdot\nabla\varepsilon\right)-\nabla^2\mathbf{E}=-\mu_0\varepsilon_0\varepsilon\frac{\partial^2\mathbf{E}}{\partial t^2}. \tag{2.2}$$

For negligible variation of the dielectric profile $\varepsilon = \varepsilon(\mathbf{r})$ over distances on the order of one optical wavelength, (2.2) simplifies to the central equation of electromagnetic wave theory,

$$\nabla^2\mathbf{E}-\frac{\varepsilon}{c^2}\frac{\partial^2\mathbf{E}}{\partial t^2}=0. \tag{2.3}$$

Practically, this equation has to be solved separately in regions of constant ε, and the obtained solutions have to been matched using appropriate boundary conditions. To cast (2.3) in a form suitable for the description of confined propagating waves, we proceed in two steps. First, we assume in all generality a harmonic time dependence $\mathbf{E}(\mathbf{r},t)=\mathbf{E}(\mathbf{r})\mathrm{e}^{-i\omega t}$ of the electric field. Inserted into (2.3), this yields

$$\nabla^2\mathbf{E}+k_0^2\varepsilon\mathbf{E}=0, \tag{2.4}$$

where $k_0 = \frac{\omega}{c}$ is the wave vector of the propagating wave in vacuum. Equation (2.4) is known as the *Helmholtz equation*.

Next, we have to define the propagation geometry. We assume for simplicity a one-dimensional problem, i.e. ε depends only on one spatial coordinate. Specifically, the waves propagate along the x-direction of a cartesian coordinate system, and show no spatial variation in the perpendicular, in-plane y-direction (see Fig. 2.1); therefore $\varepsilon = \varepsilon(z)$. Applied to electromagnetic surface problems, the plane $z = 0$ coincides with the interface sustaining the

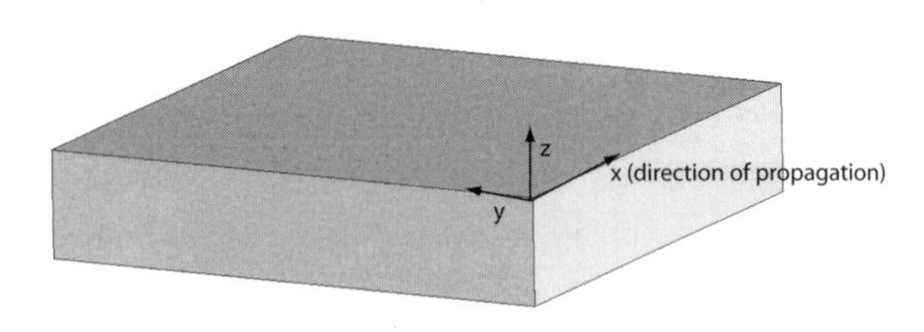

Figure 2.1. Definition of a planar waveguide geometry. The waves propagate along the x-direction in a cartesian coordinate system.

propagating waves,which can now be described as $\mathbf{E}(x, y, z) = \mathbf{E}(z)e^{i\beta x}$. The complex parameter $\beta = k_x$ is called the *propagation constant* of the traveling waves and corresponds to the component of the wave vector in the direction of propagation. Inserting this expression into (2.4) yields the desired form of the wave equation

$$\frac{\partial^2 \mathbf{E}(z)}{\partial z^2} + \left(k_0^2 \varepsilon - \beta^2\right) \mathbf{E} = 0. \tag{2.5}$$

Naturally, a similar equation exists for the magnetic field **H**.

Equation (2.5) is the starting point for the general analysis of guided electromagnetic modes in waveguides, and an extended discussion of its properties and applications can be found in [Yariv, 1997] and similar treatments of photonics and optoelectronics. In order to use the wave equation for determining the spatial field profile and dispersion of propagating waves, we now need to find explicit expressions for the different field components of **E** and **H**. This can be achieved in a straightforward way using the curl equations (1.1c, 1.1d).

For harmonic time dependence $\left(\frac{\partial}{\partial t} = -i\omega\right)$, we arrive at the following set of coupled equations

$$\frac{\partial E_z}{\partial y} - \frac{\partial E_y}{\partial z} = i\omega\mu_0 H_x \tag{2.6a}$$

$$\frac{\partial E_x}{\partial z} - \frac{\partial E_z}{\partial x} = i\omega\mu_0 H_y \tag{2.6b}$$

$$\frac{\partial E_y}{\partial x} - \frac{\partial E_x}{\partial y} = i\omega\mu_0 H_z \tag{2.6c}$$

$$\frac{\partial H_z}{\partial y} - \frac{\partial H_y}{\partial z} = -i\omega\varepsilon_0\varepsilon E_x \tag{2.6d}$$

$$\frac{\partial H_x}{\partial z} - \frac{\partial H_z}{\partial x} = -i\omega\varepsilon_0\varepsilon E_y \tag{2.6e}$$

$$\frac{\partial H_y}{\partial x} - \frac{\partial H_x}{\partial y} = -i\omega\varepsilon_0\varepsilon E_z. \tag{2.6f}$$

For propagation along the x-direction $\left(\frac{\partial}{\partial x} = i\beta\right)$ and homogeneity in the y-direction $\left(\frac{\partial}{\partial y} = 0\right)$, this system of equation simplifies to

$$\frac{\partial E_y}{\partial z} = -i\omega\mu_0 H_x \tag{2.7a}$$

$$\frac{\partial E_x}{\partial z} - i\beta E_z = i\omega\mu_0 H_y \tag{2.7b}$$

$$i\beta E_y = i\omega\mu_0 H_z \tag{2.7c}$$

$$\frac{\partial H_y}{\partial z} = i\omega\varepsilon_0\varepsilon E_x \tag{2.7d}$$

$$\frac{\partial H_x}{\partial z} - i\beta H_z = -i\omega\varepsilon_0\varepsilon E_y \tag{2.7e}$$

$$i\beta H_y = -i\omega\varepsilon_0\varepsilon E_z. \tag{2.7f}$$

It can easily be shown that this system allows two sets of self-consistent solutions with different polarization properties of the propagating waves. The first set are the transverse magnetic (TM or p) modes, where only the field components E_x, E_z and H_y are nonzero, and the second set the transverse electric (TE or s) modes, with only H_x, H_z and E_y being nonzero.

For TM modes, the system of governing equations (2.7) reduces to

$$E_x = -i\frac{1}{\omega\varepsilon_0\varepsilon}\frac{\partial H_y}{\partial z} \tag{2.8a}$$

$$E_z = -\frac{\beta}{\omega\varepsilon_0\varepsilon}H_y, \tag{2.8b}$$

and the wave equation for TM modes is

$$\frac{\partial^2 H_y}{\partial z^2} + \left(k_0^2\varepsilon - \beta^2\right) H_y = 0. \tag{2.8c}$$

For TE modes the analogous set is

$$H_x = i\frac{1}{\omega\mu_0}\frac{\partial E_y}{\partial z} \tag{2.9a}$$

$$H_z = \frac{\beta}{\omega\mu_0}E_y, \tag{2.9b}$$

with the TE wave equation

$$\frac{\partial^2 E_y}{\partial z^2} + \left(k_0^2\varepsilon - \beta^2\right) E_y = 0. \tag{2.9c}$$

With these equations at our disposal, we are now in a position to embark on the description of surface plasmon polaritons.

2.2 Surface Plasmon Polaritons at a Single Interface

The most simple geometry sustaining SPPs is that of a single, flat interface (Fig. 2.2) between a dielectric, non-absorbing half space ($z > 0$) with positive real dielectric constant ε_2 and an adjacent conducting half space ($z < 0$) described via a dielectric function $\varepsilon_1(\omega)$. The requirement of metallic character implies that $\mathrm{Re}[\varepsilon_1] < 0$. As shown in chapter 1, for metals this condition is fulfilled at frequencies below the bulk plasmon frequency ω_p. We want to look for propagating wave solutions confined to the interface, i.e. with evanescent decay in the perpendicular z-direction.

Let us first look at TM solutions. Using the equation set (2.8) in both half spaces yields

$$H_y(z) = A_2 \mathrm{e}^{i\beta x} \mathrm{e}^{-k_2 z} \tag{2.10a}$$

$$E_x(z) = i A_2 \frac{1}{\omega \varepsilon_0 \varepsilon_2} k_2 e^{i\beta x} \mathrm{e}^{-k_2 z} \tag{2.10b}$$

$$E_z(z) = -A_1 \frac{\beta}{\omega \varepsilon_0 \varepsilon_2} e^{i\beta x} \mathrm{e}^{-k_2 z} \tag{2.10c}$$

for $z > 0$ and

$$H_y(z) = A_1 \mathrm{e}^{i\beta x} \mathrm{e}^{k_1 z} \tag{2.11a}$$

$$E_x(z) = -i A_1 \frac{1}{\omega \varepsilon_0 \varepsilon_1} k_1 e^{i\beta x} \mathrm{e}^{k_1 z} \tag{2.11b}$$

$$E_z(z) = -A_1 \frac{\beta}{\omega \varepsilon_0 \varepsilon_1} e^{i\beta x} \mathrm{e}^{k_1 z} \tag{2.11c}$$

for $z < 0$. $k_i \equiv k_{z,i} (i = 1, 2)$ is the component of the wave vector perpendicular to the interface in the two media. Its reciprocal value, $\hat{z} = 1/|k_z|$, defines the evanescent decay length of the fields perpendicular to the interface,

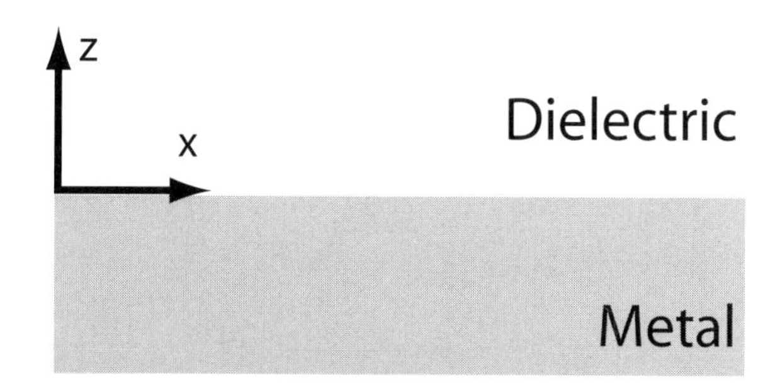

Figure 2.2. Geometry for SPP propagation at a single interface between a metal and a dielectric.

which quantifies the confinement of the wave. Continuity of H_y and $\varepsilon_i E_z$ at the interface requires that $A_1 = A_2$ and

$$\frac{k_2}{k_1} = -\frac{\varepsilon_2}{\varepsilon_1}. \tag{2.12}$$

Note that with our convention of the signs in the exponents in (2.10,2.11), confinement to the surface demands $\mathrm{Re}\,[\varepsilon_1] < 0$ if $\varepsilon_2 > 0$ - the surface waves exist only at interfaces between materials with opposite signs of the real part of their dielectric permittivities, i.e. between a conductor and an insulator. The expression for H_y further has to fulfill the wave equation (2.8c), yielding

$$k_1^2 = \beta^2 - k_0^2\varepsilon_1 \tag{2.13a}$$

$$k_2^2 = \beta^2 - k_0^2\varepsilon_2. \tag{2.13b}$$

Combining this and (2.12) we arrive at the central result of this section, the dispersion relation of SPPs propagating at the interface between the two half spaces

$$\beta = k_0\sqrt{\frac{\varepsilon_1\varepsilon_2}{\varepsilon_1 + \varepsilon_2}}. \tag{2.14}$$

This expression is valid for both real and complex ε_1, i.e. for conductors without and with attenuation.

Before discussing the properties of the dispersion relation (2.14) in more detail, we now briefly analyze the possibility of TE surface modes. Using (2.9), the respective expressions for the field components are

$$E_y(z) = A_2 \mathrm{e}^{i\beta x}\mathrm{e}^{-k_2 z} \tag{2.15a}$$

$$H_x(z) = -iA_2\frac{1}{\omega\mu_0}k_2 e^{i\beta x}\mathrm{e}^{-k_2 z} \tag{2.15b}$$

$$H_z(z) = A_2\frac{\beta}{\omega\mu_0}e^{i\beta x}\mathrm{e}^{-k_2 z} \tag{2.15c}$$

for $z > 0$ and

$$E_y(z) = A_1 \mathrm{e}^{i\beta x}\mathrm{e}^{k_1 z} \tag{2.16a}$$

$$H_x(z) = iA_1\frac{1}{\omega\mu_0}k_1 e^{i\beta x}\mathrm{e}^{k_1 z} \tag{2.16b}$$

$$H_z(z) = A_1\frac{\beta}{\omega\mu_0}e^{i\beta x}\mathrm{e}^{k_1 z} \tag{2.16c}$$

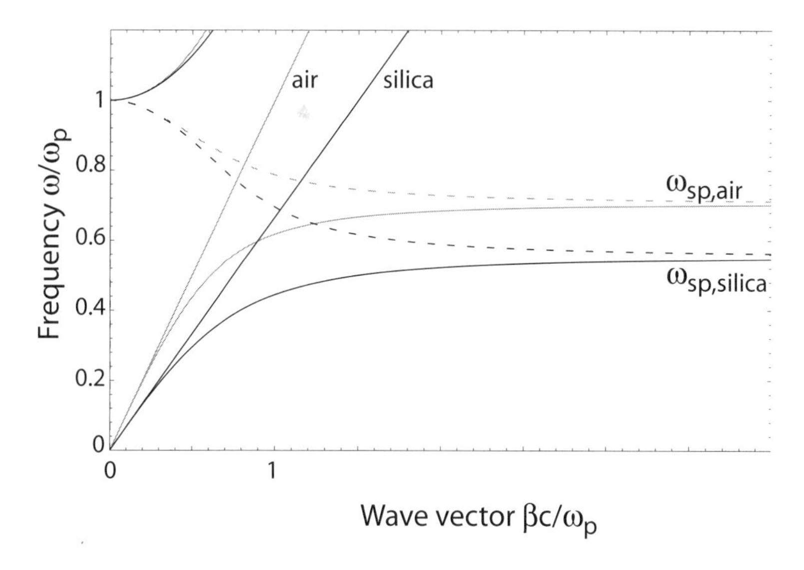

Figure 2.3. Dispersion relation of SPPs at the interface between a Drude metal with negligible collision frequency and air (gray curves) and silica (black curves).

for $z < 0$. Continuity of E_y and H_x at the interface leads to the condition

$$A_1 (k_1 + k_2) = 0. \tag{2.17}$$

Since confinement to the surface requires $\text{Re}[k_1] > 0$ and $\text{Re}[k_2] > 0$, this condition is only fulfilled if $A_1 = 0$, so that also $A_2 = A_1 = 0$. Thus, no surface modes exist for TE polarization. *Surface plasmon polaritons only exist for TM polarization.*

We now want to examine the properties of SPPs by taking a closer look at their dispersion relation. Fig. 2.3 shows plots of (2.14) for a metal with negligible damping described by the real Drude dielectric function (1.22) for an air ($\varepsilon_2 = 1$) and a fused silica ($\varepsilon_2 = 2.25$) interface. In this plot, the frequency ω is normalized to the plasma frequency ω_p, and both the real (continuous curves) and the imaginary part (broken curves) of the wave vector β are shown. Due to their bound nature, the SPP excitations correspond to the part of the dispersion curves lying to the right of the respective light lines of air and silica. Thus, special phase-matching techniques such as grating or prism coupling are required for their excitation via three-dimensional beams, which will be discussed in chapter 3. Radiation into the metal occurs in the transparency regime $\omega > \omega_p$ as mentioned in chapter 1. Between the regime of the bound and radiative modes, a frequency gap region with purely imaginary β prohibiting propagation exists.

For small wave vectors corresponding to low (mid-infrared or lower) frequencies, the SPP propagation constant is close to k_0 at the light line, and the waves extend over many wavelengths into the dielectric space. In this regime, SPPs therefore acquire the nature of a grazing-incidence light field, and are also known as *Sommerfeld-Zenneck waves* [Goubau, 1950].

In the opposite regime of large wave vectors, the frequency of the SPPs approaches the characteristic *surface plasmon frequency*

$$\omega_{\mathrm{sp}} = \frac{\omega_{\mathrm{p}}}{\sqrt{1+\varepsilon_2}}, \tag{2.18}$$

as can be shown by inserting the free-electron dielectric function (1.20) into (2.14). In the limit of negligible damping of the conduction electron oscillation (implying $\mathrm{Im}\left[\varepsilon_1(\omega)\right] = 0$), the wave vector β goes to infinity as the frequency approaches ω_{sp}, and the group velocity $v_{\mathrm{g}} \rightarrow 0$. The mode thus acquires electrostatic character, and is known as the *surface plasmon.* It can indeed be obtained via a straightforward solution of the Laplace equation $\nabla^2\phi = 0$ for the single interface geometry of Fig. 2.2, where ϕ is the electric potential. A solution that is wavelike in the x-direction and exponentially decaying in the z-direction is given by

$$\phi(z) = A_2 \mathrm{e}^{i\beta x} \mathrm{e}^{-k_2 z} \tag{2.19}$$

for $z > 0$ and

$$\phi(z) = A_1 \mathrm{e}^{i\beta x} \mathrm{e}^{k_1 z} \tag{2.20}$$

for $z < 0$. $\nabla^2\phi = 0$ requires that $k_1 = k_2 = \beta$: the exponential decay lengths $\left|\hat{z}\right| = 1/k_z$ into the dielectric and into the metal are equal. Continuity of ϕ and $\varepsilon\partial\phi/\partial z$ ensure continuity of the tangential field components and the normal components of the dielectric displacement and require that $A_1 = A_2$ and additionally

$$\varepsilon_1(\omega) + \varepsilon_2 = 0. \tag{2.21}$$

For a metal described by a dielectric function of the form (1.22), this condition is fulfilled at ω_{sp}. Comparison of (2.21) and (2.14) show that the surface plasmon is indeed the limiting form of a SPP as $\beta \rightarrow \infty$.

The above discussions of Fig. 2.3 have assumed an ideal conductor with $\mathrm{Im}\left[\varepsilon_1\right] = 0$. Excitations of the conduction electrons of real metals however suffer both from free-electron and interband damping. Therefore, $\varepsilon_1(\omega)$ is complex, and with it also the SPP propagation constant β. The traveling SPPs are damped with an energy attenuation length (also called propagation length) $L = (2\mathrm{Im}\left[\beta\right])^{-1}$, typically between 10 and 100 μm in the visible regime, depending upon the metal/dielectric configuration in question.

Fig. 2.4 shows as an example the dispersion relation of SPPs propagating at a silver/air and silver/silica interface, with the dielectric function $\varepsilon_1(\omega)$ of silver

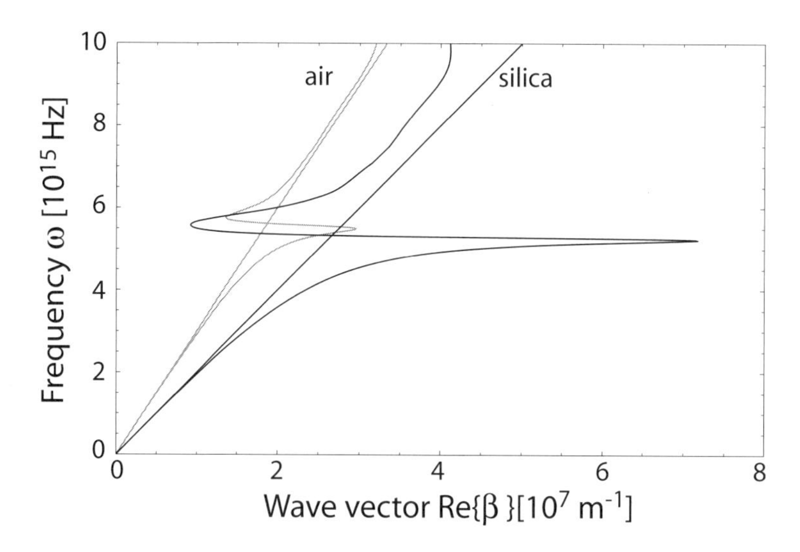

Figure 2.4. Dispersion relation of SPPs at a silver/air (gray curve) and silver/silica (black curve) interface. Due to the damping, the wave vector of the bound SPPs approaches a finite limit at the surface plasmon frequency.

taken from the data obtained by Johnson and Christy [Johnson and Christy, 1972]. Compared with the dispersion relation of completely undamped SPPs depicted in Fig. 2.3, it can be seen that the bound SPPs approach now a maximum, *finite* wave vector at the the surface plasmon frequency ω_{sp} of the system. This limitation puts a lower bound both on the wavelength $\lambda_{\mathrm{sp}} = 2\pi/\mathrm{Re}\left[\beta\right]$ of the surface plasmon and also on the amount of mode confinement perpendicular to the interface, since the SPP fields in the dielectric fall off as $e^{-|k_z||z|}$ with $k_z = \sqrt{\beta^2 - \varepsilon_2\left(\frac{\omega}{c}\right)^2}$. Also, the *quasibound*, leaky part of the dispersion relation between ω_{sp} and ω_{p} is now allowed, in contrast to the case of an ideal conductor, where $\mathrm{Re}\left[\beta\right] = 0$ in this regime (Fig. 2.3).

We finish this section by providing an example of the propagation length L and the energy confinement (quantified by $\hat{z}$) in the dielectric. As is evident from the dispersion relation, both show a strong dependence on frequency. SPPs at frequencies close to ω_{sp} exhibit large field confinement to the interface and a subsequent small propagation distance due to increased damping. Using the theoretical treatment outlined above, we see that SPPs at a silver/air interface at $\lambda_0 = 450$ nm for example have $L \approx 16\ \mu$m and $\hat{z} \approx 180$ nm. At $\lambda_0 \approx 1.5\ \mu$m however, $L \approx 1080\ \mu$m and $\hat{z} \approx 2.6\ \mu$m. The better the confinement, the lower the propagation length. This characteristic trade-off between localization and loss is typical for plasmonics. We note that field-confinement below the diffraction limit of half the wavelength in the dielectric can be achieved close to ω_{sp}. In the metal itself, the fields fall off over distances on the order of 20 nm over a wide frequency range spanning from the visible to the infrared.

2.3 Multilayer Systems

We now turn our attention to SPPs in multilayers consisting of alternating conducting and dielectric thin films. In such a system, each single interface can sustain bound SPPs. When the separation between adjacent interfaces is comparable to or smaller than the decay length $\hat{z}$ of the interface mode, interactions between SPPs give rise to coupled modes. In order to elucidate the general properties of coupled SPPs, we will focus on two specific three-layer systems of the geometry depicted in Fig. 2.5: Firstly, a thin metallic layer (I) sandwiched between two (infinitely) thick dielectric claddings (II, III), an insulator/metal/insulator (IMI) heterostructure, and secondly a thin dielectric core layer (I) sandwiched between two metallic claddings (II, III), a metal/insulator/metal (MIM) heterostructure.

Since we are here only interested in the lowest-order bound modes, we start with a general description of TM modes that are non-oscillatory in the z-direction normal to the interfaces using (2.8). For $z > a$, the field components are

$$H_y = A e^{i\beta x} e^{-k_3 z} \tag{2.22a}$$

$$E_x = iA \frac{1}{\omega \varepsilon_0 \varepsilon_3} k_3 e^{i\beta x} e^{-k_3 z} \tag{2.22b}$$

$$E_z = -A \frac{\beta}{\omega \varepsilon_0 \varepsilon_3} e^{i\beta x} e^{-k_3 z}, \tag{2.22c}$$

while for $z < -a$ we get

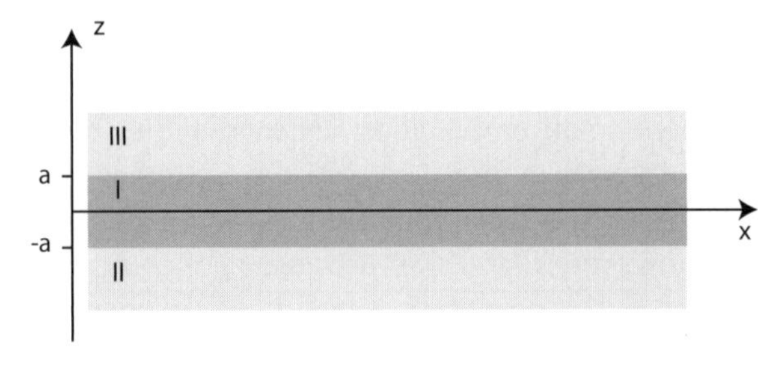

Figure 2.5. Geometry of a three-layer system consisting of a thin layer I sandwiched between two infinite half spaces II and III.

$$H_y = B\mathrm{e}^{i\beta x}\mathrm{e}^{k_2 z} \tag{2.23a}$$

$$E_x = -iB\frac{1}{\omega\varepsilon_0\varepsilon_2}k_2\mathrm{e}^{i\beta x}\mathrm{e}^{k_2 z} \tag{2.23b}$$

$$E_z = -B\frac{\beta}{\omega\varepsilon_0\varepsilon_2}\mathrm{e}^{i\beta x}\mathrm{e}^{k_2 z}. \tag{2.23c}$$

Thus, we demand that the fields decay exponentially in the claddings (II) and (III). Note that for simplicity as before we denote the component of the wave vector perpendicular to the interfaces simply as $k_i \equiv k_{z,i}$.

In the core region $-a < z < a$, the modes localized at the bottom and top interface couple, yielding

$$H_y = C\mathrm{e}^{i\beta x}\mathrm{e}^{k_1 z} + D\mathrm{e}^{i\beta x}\mathrm{e}^{-k_1 z} \tag{2.24a}$$

$$E_x = -iC\frac{1}{\omega\varepsilon_0\varepsilon_1}k_1\mathrm{e}^{i\beta x}\mathrm{e}^{k_1 z} + iD\frac{1}{\omega\varepsilon_0\varepsilon_1}k_1\mathrm{e}^{i\beta x}\mathrm{e}^{-k_1 z} \tag{2.24b}$$

$$E_z = C\frac{\beta}{\omega\varepsilon_0\varepsilon_1}\mathrm{e}^{i\beta x}\mathrm{e}^{k_1 z} + D\frac{\beta}{\omega\varepsilon_0\varepsilon_1}\mathrm{e}^{i\beta x}\mathrm{e}^{-k_1 z}. \tag{2.24c}$$

The requirement of continutity of H_y and E_x leads to

$$A\mathrm{e}^{-k_3 a} = C\mathrm{e}^{k_1 a} + D\mathrm{e}^{-k_1 a} \tag{2.25a}$$

$$\frac{A}{\varepsilon_3}k_3\mathrm{e}^{-k_3 a} = -\frac{C}{\varepsilon_1}k_1\mathrm{e}^{k_1 a} + \frac{D}{\varepsilon_1}k_1\mathrm{e}^{-k_1 a} \tag{2.25b}$$

at $z = a$ and

$$B\mathrm{e}^{-k_2 a} = C\mathrm{e}^{-k_1 a} + D\mathrm{e}^{k_1 a} \tag{2.26a}$$

$$-\frac{B}{\varepsilon_2}k_2\mathrm{e}^{-k_2 a} = -\frac{C}{\varepsilon_1}k_1\mathrm{e}^{-k_1 a} + \frac{D}{\varepsilon_1}k_1\mathrm{e}^{k_1 a} \tag{2.26b}$$

at $z = -a$, a linear system of four coupled equations. H_y further has to fulfill the wave equation (2.8c) in the three distinct regions, via

$$k_i^2 = \beta^2 - k_0^2\varepsilon_i \tag{2.27}$$

for $i = 1, 2, 3$. Solving this system of linear equations results in an implicit expression for the dispersion relation linking β and ω via

$$\mathrm{e}^{-4k_1 a} = \frac{k_1/\varepsilon_1 + k_2/\varepsilon_2}{k_1/\varepsilon_1 - k_2/\varepsilon_2}\frac{k_1/\varepsilon_1 + k_3/\varepsilon_3}{k_1/\varepsilon_1 - k_3/\varepsilon_3}. \tag{2.28}$$

We note that for infinite thickness ($a \rightarrow \infty$), (2.28) reduces to (2.12), the equation of two uncoupled SPP at the respective interfaces.

We will from this point onwards consider the interesting special case where the sub- and the superstrates (II) and (III) are equal in terms of their dielectric response, i.e. $\varepsilon_2 = \varepsilon_3$ and thus $k_2 = k_3$. In this case, the dispersion relation (2.28) can be split into a pair of equations, namely

$$\tanh k_1 a = -\frac{k_2 \varepsilon_1}{k_1 \varepsilon_2} \tag{2.29a}$$

$$\tanh k_1 a = -\frac{k_1 \varepsilon_2}{k_2 \varepsilon_1}. \tag{2.29b}$$

It can be shown that equation (2.29a) describes modes of odd vector parity ($E_x(z)$ is odd, $H_y(z)$ and $E_z(z)$ are even functions), while (2.29b) describes modes of even vector parity ($E_x(z)$ is even function, $H_y(z)$ and $E_z(z)$ are odd).

The dispersion relations (2.29a, 2.29b) can now be applied to IMI and MIM structures to investigate the properties of the coupled SPP modes in these two systems. We first start with the IMI geometry - a thin metallic film of thickness $2a$ sandwiched between two insulating layers. In this case $\varepsilon_1 = \varepsilon_1(\omega)$ represents the dielectric function of the metal, and ε_2 the positive, real dielectric constant of the insulating sub- and superstrates. As an example, Fig. 2.6 shows the dispersion relations of the odd and even modes (2.29a, 2.29b) for an air/silver/air geometry for two different thicknesses of the silver thin film. For

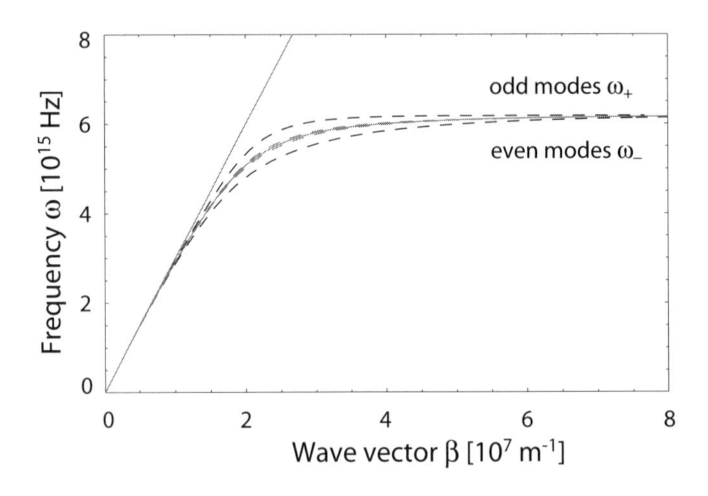

Figure 2.6. Dispersion relation of the coupled odd and even modes for an air/silver/air multilayer with a metal core of thickness 100 nm (dashed gray curves) and 50 nm (dashed black curves). Also shown is the dispersion of a single silver/air interface (gray curve). Silver is modeled as a Drude metal with negligible damping.

simplicity, here the dielectric function of silver is approximated via a Drude model with negligible damping ($\varepsilon(\omega)$ real and of the form (1.22)), so that $\mathrm{Im}\left[\beta\right] = 0$.

As can be seen, the odd modes have frequencies ω_+ higher than the respective frequencies for a single interface SPP, and the even modes lower frequencies ω_-. For large wave vectors β (which are only achievable if $\mathrm{Im}\left[\varepsilon(\omega)\right] = 0$), the limiting frequencies are

$$\omega_+ = \frac{\omega_\mathrm{p}}{\sqrt{1+\varepsilon_2}}\sqrt{1 + \frac{2\varepsilon_2 \mathrm{e}^{-2\beta a}}{1+\varepsilon_2}} \tag{2.30a}$$

$$\omega_- = \frac{\omega_\mathrm{p}}{\sqrt{1+\varepsilon_2}}\sqrt{1 - \frac{2\varepsilon_2 \mathrm{e}^{-2\beta a}}{1+\varepsilon_2}}. \tag{2.30b}$$

Odd modes have the interesting property that upon decreasing metal film thickness, the confinement of the coupled SPP to the metal film decreases as the mode evolves into a plane wave supported by the homogeneous dielectric environment. For real, absorptive metals described via a complex $\varepsilon(\omega)$, this implies a drastically increased SPP propagation length [Sarid, 1981]. These *long-ranging* SPPs will be further discussed in chapter 7. The even modes exhibit the opposite behavior - their confinement to the metal increases with decreasing metal film thickness, resulting in a reduction in propagation length.

Moving on to MIM geometries, we now set $\varepsilon_2 = \varepsilon_2(\omega)$ as the dielectric function of the metal and ε_1 as the dielectric constant of the insulating core in equations (2.29a, 2.29b). From an energy confinement point of view, the most interesting mode is the fundamental odd mode of the system, which does not exhibit a cut-off for vanishing core layer thickness [Prade et al., 1991]. Fig. 2.7 shows the dispersion relation of this mode for a silver/air/silver heterostructure. This time, the dielectric function $\varepsilon(\omega)$ was taken as a complex fit to the dielectric data of silver obtained by Johnson and Christy [Johnson and Christy, 1972]. Thus β does not go to infinity as the surface plasmon frequency is approached, but folds back and eventually crosses the light line, as for SPPs propagating at single interfaces.

It is apparent that large propagation constants β can be achieved even for excitation well below ω_{sp}, provided that the width of the dielectric core is chosen sufficiently small. The ability to access such large wave vectors and thus small penetration lengths $\hat{z}$ into the metallic layers by adjusting the geometry indicates that localization effects that for a single interface can only be sustained at excitations near ω_{sp}, can for such MIM structures also be attained for excitation out in the the infrared. An analysis of various other MIM structures, for example concentric shells, has given similar results [Takahara et al., 1997]. Geometries amendable to easy fabrication such as triangular metal V-grooves

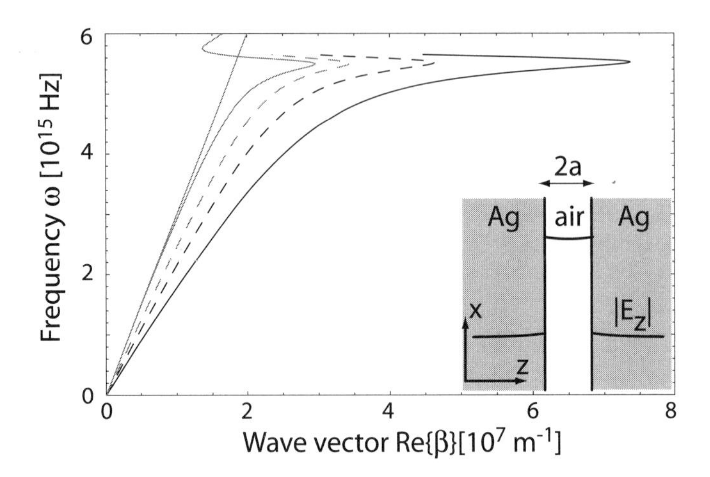

Figure 2.7. Dispersion relation of the fundamental coupled SPP modes of a silver/air/silver multilayer geometry for an air core of size 100 nm (broken gray curve), 50 nm (broken black curve), and 25 nm (continuous black curve). Also shown is the dispersion of a SPP at a single silver/air interface (gray curve) and the air light line (gray line).

on a flat metal surface have already shown great promise for applications in waveguiding, which will be presented in chapter 7.

We have limited our discussion of coupled SPPs in three-layer structures to the fundamental bound modes of the system, with a view on applications in waveguiding and confinement of electromagnetic energy. It is important to note that the family of modes supported by this geometry is much richer than described in this treatment. For example, for IMI structures, we have omitted a discussion of leaky modes, and MIM layers can also exhibit oscillatory modes for sufficient thickness of the dielectric core. Additionally, the coupling between SPPs at the two core/cladding interfaces changes significantly when the dielectric constants of the sub- and superstrates are different, so that $\varepsilon_2 \neq \varepsilon_3$, prohibiting phase-matching between the modes located at the two interfaces. A detailed treatment of these cases can be found in [Economou, 1969, Burke and Stegeman, 1986, Prade et al., 1991].

2.4 Energy Confinement and the Effective Mode Length

In chapter 5 we will see that using localized surface plasmons in metal nanoparticles, electromagnetic energy can be confined or squeezed into volumes smaller than the diffraction limit $(\lambda_0/2n)^3$, where $n = \sqrt{\varepsilon}$ is the refractive index of the surrounding medium. This high confinement leads to a concomitant field enhancement and is of prime importance in plasmonics, enabling a great variety of applications in optical sensing, as will be discussed in chapter 9. In the essentially one-dimensional cases of single interfaces and

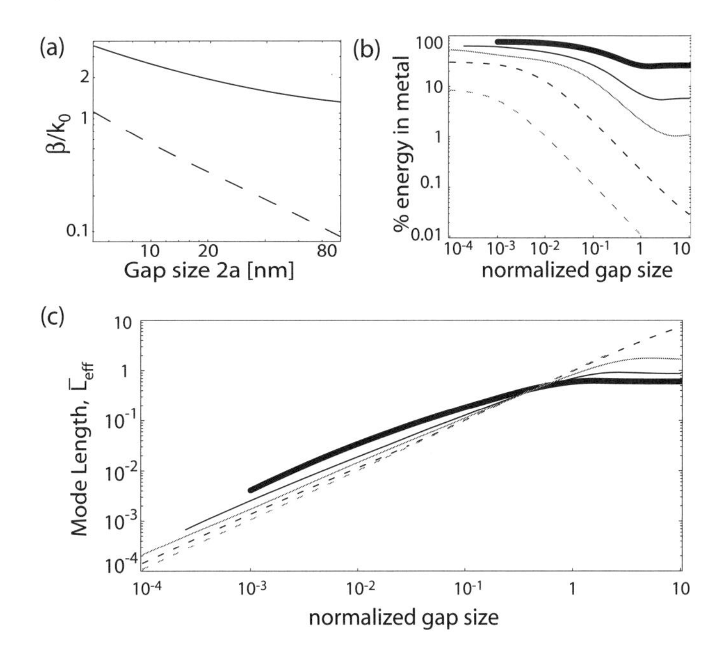

Figure 2.8. Energy confinement in a gold/air/gold MIM structure. (a) Real (solid curve) and imaginary (dashed curve) part of the normalized propagation constant β versus gap size at $\lambda_0 = 850$nm. (b) Fraction of electric field energy residing inside the metallic half spaces as a function of normalized gap size for excitation at $\lambda_0 = 600$ nm (thick curve), 850nm (black curve), 1.5 μm (gray curve), 10 μm (broken black curve), and 100 μm (broken gray curve). (c) Effective mode length L_{eff} normalized to free-space wavelength λ_0 as a function of gap size. Adapted from [Maier, 2006b].

multilayer structures presented above that support propagating SPPs, energy localization below the diffraction limit perpendicular to the interface(s) is also possible. We have already hinted at this phenomenon when stating that the field decay length $\hat{z}$ in the dielectric layers can be significantly smaller than λ_0/n.

However, care must be taken when quantifying energy confinement, since a sub-wavelength field decay length $\hat{z}$ on the dielectric side of the interface implies that a significant amount of the total electric field energy of the SPP mode resides inside the metal. This energy must be taken into account using (1.38) when calculating the spatial distribution of the electric energy density, since for the quantification of the strength of interactions between light and matter (e.g. a molecule placed into the field), the field strength per unit energy (i.e., single photon) is of importance.

Taking a gold/air/gold MIM heterostructure as an example, Fig. 2.8(a) shows the evolution of both the real and imaginary parts of the propagation constant β of the fundamental SPP mode with varying gap size for excitation at a free space wavelength of $\lambda_0 = 850$ nm, calculated using Drude fits to the dielectric function of gold [Johnson and Christy, 1972, Ordal et al., 1983]. Both parts increase with decreasing gap size, since the mode is becoming more electron-plasma in character, suggesting that the electromagnetic energy is residing increasingly in the metal half-spaces. A plot of the fractional amount of the electric field energy inside the metal regions is shown in Fig. 2.8(b) for excitation at wavelengths $\lambda_0 = 600$ nm, 850 nm, 1.5 μm, 10 μm, and 100 μm (= 3 THz). For a gap of 20 nm for example, at $\lambda_0 = 850$ nm this fraction already reaches 40%. Note that the gap size is normalized to the respective free space wavelength. It is apparent that along with the increased localization of the field to the gold/air interface, either via small gap sizes or excitation closer to ω_{sp}, comes a shift of the energy into the metal regions.

In order to get a better handle on the consequences of increasing fractions of the total energy of the mode entering the metallic cladding upon decreasing size of the dielectric gap, we can define in analogy to the effective mode *volume* V_{eff} used to quantify the strength of light-matter interactions in cavity quantum electrodynamics [Andreani et al., 1999] an effective mode *length* L_{eff}, with

$$L_{eff}(z_0)u_{eff}(z_0) = \int u_{eff}(z)dz. \tag{2.31}$$

$u_{eff}(z_0)$ represents the electric field energy density at a position z_0 of interest within the air core (e.g. the location of an emitter). In this one-dimensional picture, the effective mode length is therefore given as the ratio of the total energy of the SPP mode divided by the energy density (energy per unit length) at the position of interest, which is often taken as the position of highest field. In a quantized picture for normalized total energy, the inverse of the effective mode length thus quantifies the field strength per single SPP excitation. More details can be found in [Maier, 2006b].

A determination of the effective mode length of MIM structures allows an examination how the electric field strength per SPP excitation in the air gap scales as a function of the gap size. Fig. 2.8(c) shows the variation of $\bar{L}_{eff}$ (normalized to the free-space wavelength λ_0) with normalized gap size. z_0 is taken to be at the air side of the air/gold boundary, where the electric field strength is maximum. The mode lengths drop well below $\lambda_0/2$, demonstrating that plasmonic metal structures can indeed sustain *effective* as well as *physical* mode lengths below the diffraction limit of light. The trend in L_{eff} with gap size tends to scale with the physical extent of the air gap. For large normalized gap sizes and low frequencies, this is due to the delocalized nature of the surface plasmon, leading to smaller mode lengths for excitation closer to the surface plasmon frequency ω_{sp} for the same normalized gap size.

As the gap size is reduced to a point where the dispersion curve of the SPP mode turns over (see Fig. 2.7) and energy begins to enter the metallic half spaces, the continued reduction in mode length is due to an increase in field localization to the metal-air interface. In this regime, excitations with lower frequencies show smaller mode lengths for the same normalized gap size than excitations closer to the plasmon resonance, due to the fact that more energy resides inside the metal for the latter. We note that for very small gaps with $2a < 2$ nm, the effects of local fields due to unscreened surface electrons become important [Larkin et al., 2004], leading to a further decrease in L_{eff}. This cannot be captured using the dielectric function approach.

To summarize, we see that despite the penetration of a significant amount of energy of a SPP mode into the conducting medium (for excitation near ω_{sp} or in small gap structures), the associated large propagation constants β ensure that the effective extent of the mode perpendicular to the interface(s) drops well below the diffraction limit.

Chapter 3

EXCITATION OF SURFACE PLASMON POLARITONS AT PLANAR INTERFACES

Surface plasmon polaritons propagating at the flat interface between a conductor and a dielectric are essentially two-dimensional electromagnetic waves. Confinement is achieved since the propagation constant β is greater than the wave vector k in the dielectric, leading to evanescent decay on both sides of the interface. The SPP dispersion curve therefore lies to the right of the light line of the dielectric (given by $\omega = ck$), and excitation by three-dimensional light beams is not possible unless special techniques for phase-matching are employed. Alternatively, thin film geometries such as insulator-metal-insulator heterostructures sustaining weakly confined SPPs are amenable to end-fire coupling, relying on spatial mode-matching rather than phase-matching.

This chapter reviews the most common techniques for SPP excitation. After a discussion of excitation using charged particles, various optical techniques for phase-matching such as prism and grating coupling as well as excitation using highly focused beams will be presented. Wave vectors in excess of $|\mathbf{k}|$ can also be achieved using illumination in the near-field, making use of evanescent waves in the immediate vicinity of a sub-wavelength aperture. The chapter closes with a brief look at the excitation of SPPs in nanoparticle waveguides and multilayer structures using optical fiber tapers or end-fire excitation. This allows coupling of SPPs to modes in conventional dielectric waveguides. Techniques for the excitation and investigation of localized plasmons in metallic nanostructures such as various forms of microscopy and cathodoluminescence will be presented in chapter 10.

3.1 Excitation upon Charged Particle Impact

Surface plasmons - the non-propagating, quasi-static electromagnetic surfaces modes at ω_{sp} described by (2.21) - were theoretically investigated by Ritchie in the context of loss spectra of low-energy electron beams undergoing

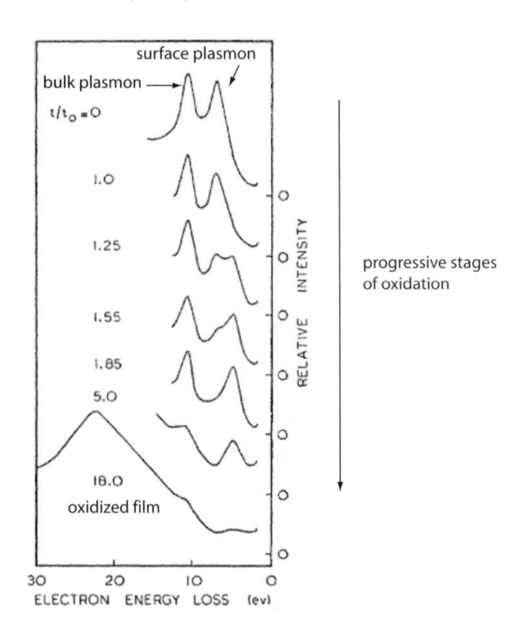

Figure 3.1. Electron energy loss spectra of a thin magnesium film in progressive stages of oxidation. Reprinted with permission from [Powell and Swan, 1960]. Copyright 1960 by the American Physical Society.

diffraction at thin metallic films [Ritchie, 1957]. Apart from the expected volume plasmon excitation of energy $\hbar\omega_{\text{p}}$, this study predicted an additional loss at a lower energy $\hbar\omega_{\text{p}}/\sqrt{2}$, subsequently termed *low-lying energy loss*. While loss spectroscopy of electron diffraction at metal films was traditionally employed for the excitation of longitudinal volume plasmons, Powell and Swan observed the additional peak in electron energy loss spectra of magnesium and aluminum in reflection (Fig. 3.1) [Powell and Swan, 1960]. A shift of the peak to lower energies during oxidation of the metal films suggested it being associated with an electromagnetic excitation at the metal/air surface, which during the experiment was slowly evolving into a metal/oxide interface.

The energy loss at $\hbar\omega_{\text{p}}/\sqrt{2}$ indeed turned out to be due to the surface excitation previously predicted by Ritchie for a metal/air interface. It corresponds to the surface plasmon excitation described in the previous chapter. Subsequent theoretical investigations of surface plasmon waves in the context of electron loss spectroscopy confirmed the $\omega_{\text{sp}} = \frac{\omega_{\text{p}}}{\sqrt{1+\varepsilon}}$ dependence of the resonance frequency on the dielectric coating (explaining the influence of an oxide layer), and the possibility of even and odd coupled modes akin to (2.29) sustained by thin metallic films [Stern and Ferrell, 1960].

While low-energy electron *diffraction* experiments can only detect excitations at the asymptotic surface plasmon energy $\hbar\omega_{\text{sp}}$, an analysis of the change

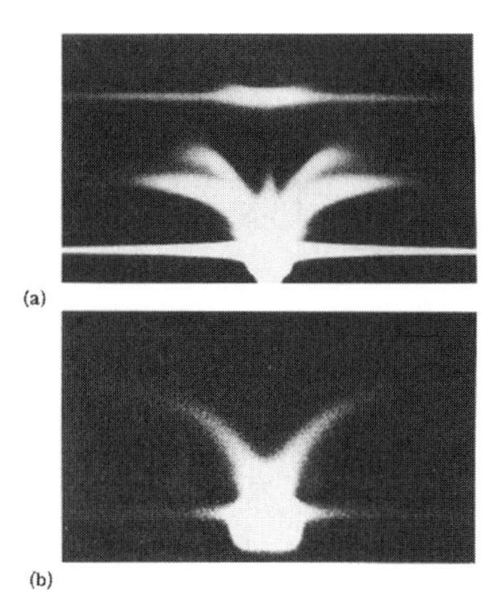

Figure 3.2. Direct map of the SPP dispersion formed via energy-loss spectra for transmission of a 75-keV electron beam through a 16 nm aluminum film at normal incidence. Exposure times are 15 minutes (a) and 3 minutes (b). Reprinted with permission from [Pettit et al., 1975]. Copyright 1975 by the American Physical Society.

in energy and momentum of fast electrons *transmitted* through thin metal films enables a study of the full dispersion relation of SPPs, as long as the angular divergence of the beam is low. Using this method, the dispersion of SPPs, including the radiative branch above ω_p, was analyzed in a number of early studies [Vincent and Silcox, 1973, Pettit et al., 1975]. For example, Pettit and co-workers demonstrated the splitting of the SPP mode into even and odd modes in a thin (16 nm) oxidized aluminum film by studying the transmission

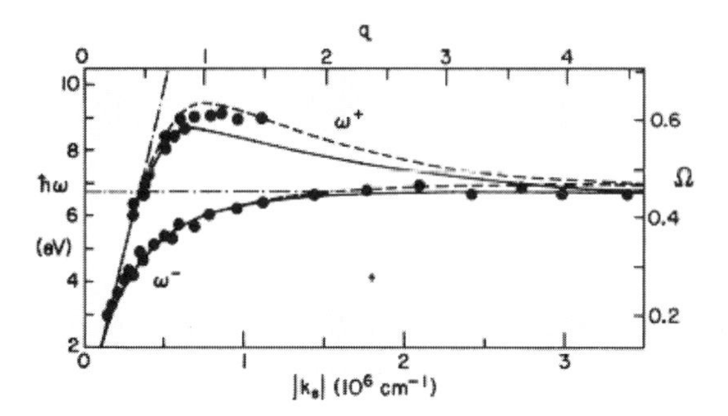

Figure 3.3. Comparison of the experimental data presented in Fig. 3.2(dots) with the theoretical dispersion curves of the two coupled modes. For the theoretical analysis see Fig. 2.6. For the calculations, the aluminum film has been assumed to be embedded into amorphous alumina (dashed curves) or alumina in its α-phase (continuous curves). Reprinted with permission from [Pettit et al., 1975]. Copyright 1975 by the American Physical Society.

of a 75-keV electron beam [Pettit et al., 1975]. Using a Wien filter spectrometer, a direct image of the dispersion relations could be obtained, shown in Fig. 3.2. The bright central spots correspond to undeflected electrons, and the two horizontal lines to volume plasmon excitations (upper line) and phonon and elastic scattering (lower lines). Additionally, the characteristic dispersion of the high- and low-frequency modes ω_+ and ω_- is clearly visible, and compares favorably with a theoretical study of the thin film (Fig. 3.3).

3.2 Prism Coupling

Surface plasmon polaritons on a flat metal/dielectric interface cannot be excited directly by light beams since $\beta > k$, where k is the wave vector of light on the dielectric side of the interface. Therefore, the projection along the interface of the momentum $k_x = k \sin\theta$ of photons impinging under an angle θ to the surface normal is always smaller than the SPP propagation constant β, even at grazing incidence, prohibiting phase-matching. We have already expanded on this fact when noting that the SPP dispersion curve (2.14) lies outside the light cone of the dielectric.

However, phase-matching to SPPs can be achieved in a three-layer system consisting of a thin metal film sandwitched between two insulators of different dielectric constants. For simplicity, we will take one of the insulators to be air ($\varepsilon = 1$). A beam reflected at the interface between the insulator of higher dielectric constant ε, usually in the form of a prism (see Fig. 3.4), and the metal will have an in-plane momentum $k_x = k\sqrt{\varepsilon} \sin\theta$, which is sufficient to excite SPPs at the interface between the metal and the lower-index dielectric, i.e. in this case at the metal/air interface. This way, SPPs with propagation constants β between the light lines of air and the higher-index dielectric can be excited (Fig. 3.5). SPP excitation manifests itself as a minimum in the reflected beam intensity. Note that phase-matching to SPPs at the prism/metal interface cannot be achieved, since the respective SPP dispersion lies outside the prism light cone (Fig. 3.5).

This coupling scheme - also known as attenuated total internal reflection - therefore involves tunneling of the fields of the excitation beam to the metal/air

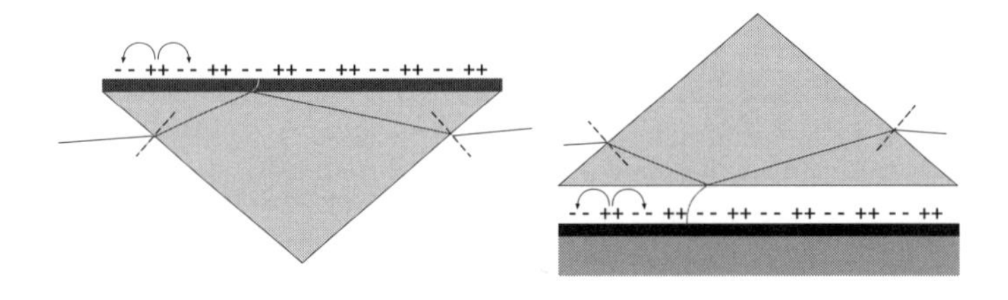

Figure 3.4. Prism coupling to SPPs using attenuated total internal reflection in the Kretschmann (left) and Otto (right) configuration. Also drawn are possible lightpaths for excitation.

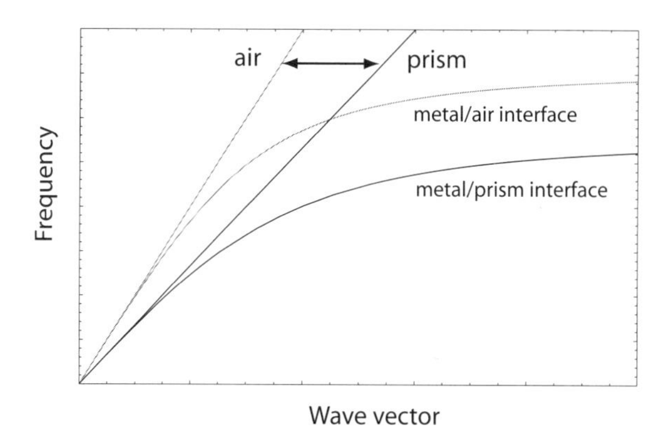

Figure 3.5. Prism coupling and SPP dispersion. Only propagation constants between the light lines of air and the prism (usually glass) are accessible, resulting in additional SPP damping due to leakage radiation into the latter: the excited SPPs have propagation constants *inside* the prism light cone.

interface where SPP excitation takes place. Two different geometries for prism coupling are possible, depicted in Fig. 3.4. The most common configuration is the Kretschmann method [Kretschmann and Raether, 1968], in which a thin metal film is evaporated on top of a glass prism. Photons from a beam impinging from the glass side at an angle greater than the critical angle of total internal reflection tunnel through the metal film and excite SPPs at the metal/air interface. Another geometry is the Otto configuration [Otto, 1968], in which the prism is separated from the metal film by a thin air gap. Total internal reflection takes place at the prism/air interface, exciting SPPs via tunneling to the air/metal interface. This configuration is preferable when direct contact with the metal surface is undesirable, for example for studies of surface quality.

We want to stress that SPPs excited using phase-matching via $\beta = k\sqrt{\varepsilon}\sin\theta$ are inherently *leaky waves*, i.e. they lose energy not only due to the inherent absorption inside the metal, but also due to leakage of radiation into the prism: the excited propagation constants lie within the prism light cone (Fig. 3.5). The minimum in the intensity of the reflected beam is due to destructive interference between this leakage radiation and the reflected part of the excitation beam. For an optimum metal film thickness, the destructive interference can be perfect, providing a zero in the reflected beam intensity, so that leakage radiation cannot be detected.

Using an analysis of this system based on the Fresnel equations [Kretschmann, 1971, Raether, 1988], it can be shown that this optimum case is achieved if the damping Γ_{LR} due to leakage radiation is equal to the damping Γ_{abs} due to absorption (critical coupling). $\Gamma_{\mathrm{abs}} = \mathrm{Im}\left[\beta_0\right]$, where β_0 is the SPP propagation constant of the single interface calculated via (2.14). For a metal layer with a dielectric function $\varepsilon_1(\omega)$ fulfilling $|\text{Re}[\varepsilon_1]| \gg 1$ and $|\text{Im}[\varepsilon_1]| \ll |\text{Re}[\varepsilon_1]|$, the reflection coefficient can be approximated via the Lorentzian

$$R = 1 - \frac{4\Gamma_{\text{LR}}\Gamma_{\text{abs}}}{\left[\beta - (\beta_0 + \Delta\beta)\right]^2 + (\Gamma_{\text{LK}} + \Gamma_{\text{abs}})^2}. \tag{3.1}$$

It is apparent that the SPP propagation constant β of the prism/metal/air system is shifted by an amount $\left|\text{Re}\left[\Delta\beta\right]\right|$ from the single interface value β_0. The imaginary part $\text{Im}\left[\Delta\beta\right] \equiv \Gamma_{\text{LK}}$ describes the contribution of radiation damping to the total loss. $\Delta\beta$ can be expressed via a calculation of the Fresnel reflection coefficients and depends on the thickness of the metal layer [Kretschmann, 1971, Raether, 1988].

The prism coupling technique is also suitable for exciting coupled SPP modes in MIM or IMI three-layer systems. By using appropriate index-matching oils, both the long-ranging high frequency mode ω_+ and the low frequency mode ω_- of higher attenuation have been excited for oil/silver/silica and also oil/aluminum/silica IMI structures brought into contact with a prism [Quail et al., 1983]. For the long-ranging mode, a reduction of the angular spread of the reflection minimum by an order of magnitude compared to the uncoupled mode at a single interface has been confirmed. This sharpening of the resonant feature is due to the decreased amount of energy in the metal film and thus decreased attenuation of the coupled SPP.

3.3 Grating Coupling

The mismatch in wave vector between the in-plane momentom $k_x = k \sin\theta$ of impinging photons and β can also be overcome by patterning the metal surface with a shallow grating of grooves or holes with lattice constant a. For the simple one-dimensional grating of grooves depicted in Fig. 3.6, phase-matching takes place whenever the condition

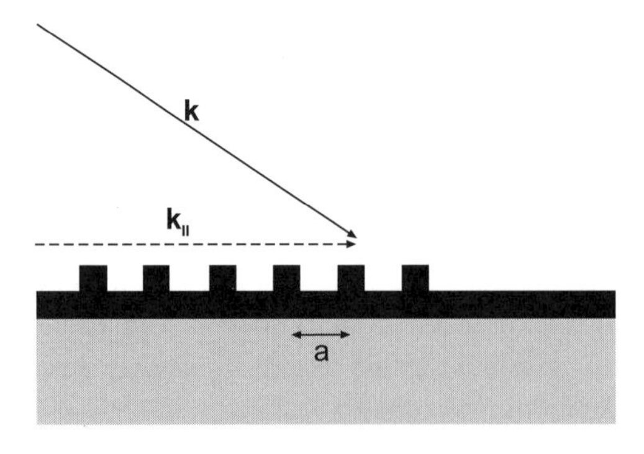

Figure 3.6. Phase-matching of light to SPPs using a grating.

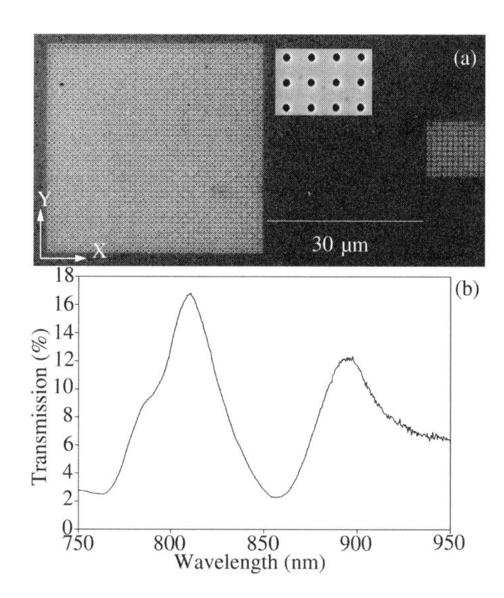

Figure 3.7. (a) SEM image of two microhole arrays with period 760 nm and hole diameter 250 nm separated by 30 μm used for sourcing (right array) and probing (left array) of SPPs. The inset shows a close-up of individual holes. (b) Normal-incidence white light transmission spectrum of the arrays. Reprinted with permission from [Devaux et al., 2003]. Copyright 2003, American Institute of Physics.

$$\beta = k \sin\theta \pm \nu g \tag{3.2}$$

is fulfilled, where $g = \frac{2\pi}{a}$ is the reciprocal vector of the grating, and $\nu = (1, 2, 3 \ldots)$. As with prism coupling, excitation of SPPs is detected as a minimum in the reflected light.

The reverse process can also take place: SPPs propagating along a surface modulated with a grating can couple to light and thus radiate. The gratings need not be milled directly into the metal surface, but can also consist of dielectric material. For example, Park and co-workers have demonstrated out-coupling of SPPs using a dielectric grating of a depth of only several nanometres with an efficiency of about 50% [Park et al., 2003]. By designing the shape of the grating, the propagation direction of SPPs can be influenced and even focusing can be achieved, which was shown by Offerhaus and colleagues using noncollinear phase-matching [Offerhaus et al., 2005]. Some studies of manipulation of SPP propagation using modulated surfaces will be presented in chapter 7 on waveguiding.

As an example of SPP excitation and their decoupling via gratings, Fig. 3.7a shows a scanning electron microscopy (SEM) image of a flat metal film patterned with two arrays of sub-wavelength holes [Devaux et al., 2003]. In this study, the small array on the right was used for the excitation of SPPs via a

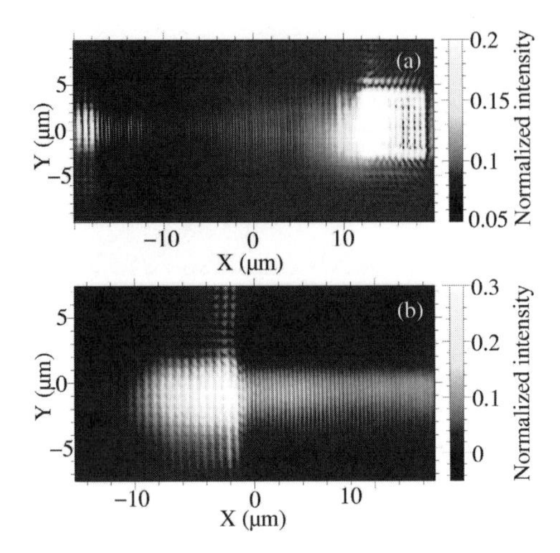

Figure 3.8. (a) Near-field optical image of the pattern presented in Fig. 3.7 when the illuminating laser is focused on the small array on the right with the electric field polarised in the x-direction. (b) Detail of image (a) showing propagating SPPs and the edge of the left outcoupling array. A wavelength $\lambda = 800$ nm was chosen so as to coincide with the airside transmission peak in Fig. 3.7. Reprinted with permission from [Devaux et al., 2003]. Copyright 2003, American Institute of Physics.

normally-incident beam, while the larger array on the left decoupled the propagating SPPs to the radiation continuum. The wavelengths of phase-matching are revealed via a normal-incidence transmission spectrum, with in this case yielded a peak at $\lambda = 815$ nm due to excitation of a SPP mode at the metal/air interface (Fig. 3.7b). Near-field optical images of the excitation and detection region as well as of the propagating SPPs are shown in Fig. 3.8. The streak between the two arrays corresponds to the propagating SPPs, showing rapid attenuation as the left hole array used for decoupling is encountered.

For one-dimensional gratings, significant changes to the SPP dispersion relation occur if the gratings are sufficiently deep so that the modulation can no longer be treated as a small perturbation of the flat interface. Appreciable band gaps appear already for a groove depth on the order of 20 nm for metallic gratings. For even larger depths, localized modes inside the grooves lead to distortions of the first higher-order band folded back at the Brillouin zone boundary, enabling coupling even for short pitches $a < \lambda/2$ upon normal incidence due to a lowering in frequency of the modified SPP dispersion curve. For more details on these effects we refer to the study by Hooper and Sambles [Hooper and Sambles, 2002]. The influence of surface structure on the dispersion of SPPs will also be further elucidated in chapter 6 on SPPs at lower frequencies.

More generally, SPPs can also be excited on films in areas with random surface roughness or manufactured localized scatterers. Momentum components Δk_x are provided via scattering, so that the phase-matching condition

$$\beta = k \sin\theta \pm \Delta k_x \tag{3.3}$$

can be fulfilled. The efficiency of coupling can be assessed by for example measuring the leakage radiation into a glass prism situated underneath the metal film, which was demonstrated by Ditlbacher and co-workers for a flat film with a small number of ridges to couple a normal-incidence beam to propagating SPPs [Ditlbacher et al., 2002a]. We note that (3.3) implies that random surface roughness also constitutes an additional loss channel for SPP propagation via coupling to radiation.

3.4 Excitation Using Highly Focused Optical Beams

As a variant of the traditional prism coupling technique described in section 3.2, a microscope objective of high numerical aperture can be used for SPP excitation. Fig 3.9 shows a typical setup [Bouhelier and Wiederrecht, 2005]. An oil-immersion objective is brought into contact with the glass substrate (of refractive index n) of a thin metal film via a layer of index-matched immersion oil. The high numerical aperture of the objective ensures a large angular spread of the focused excitation beam, including angles $\theta > \theta_c$ greater than the critical angle of total internal reflection at a glass/air interface.

This way, wave vectors $k_x = \beta$ are available for phase-matching to SPPs at the metal/air interface at the corresponding angle $\theta_{SPP} = \arcsin(\beta/nk_0) > \theta_c$. Off-axis entrance of the excitation beam into the objective can further ensure an intensity distribution preferentially around θ_{SPP}, thus reducing the amount

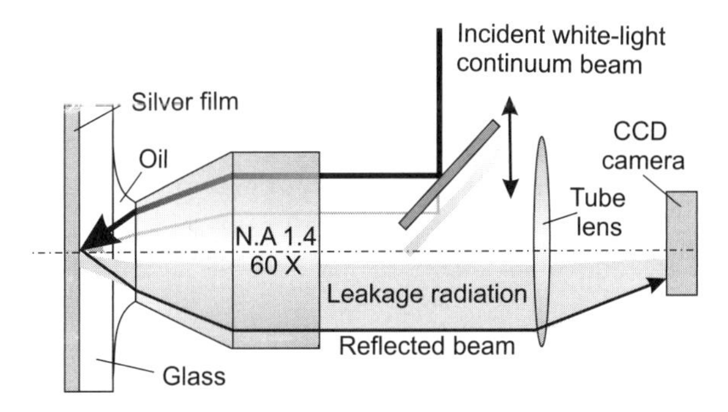

Figure 3.9. Schematic of the excitation of a white-light continuum of SPPs and their observation via detection of the leakage radiation using an index-matched oil immersion lens. Reprinted with permission from [Bouhelier and Wiederrecht, 2005]. Copyright 2005 by the Optical Society of America.

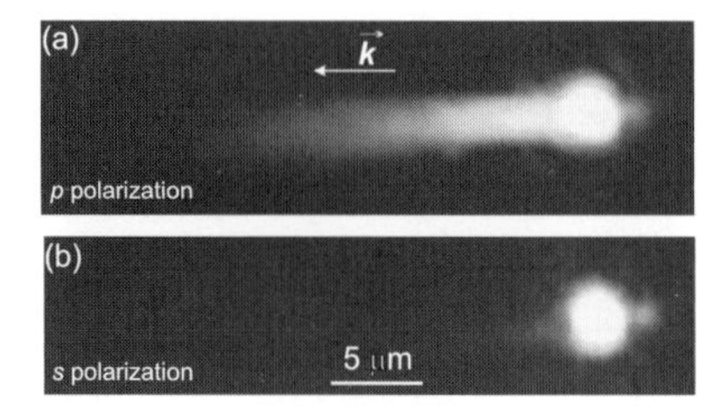

Figure 3.10. (a) Leakage radiation intensity distribution for a TM polarized white-light continuum excitation beam, showing SPPs propagating away from the excitation spot. (b) No SPP excitation is observed for TE polarization. Reprinted with permission from [Bouhelier and Wiederrecht, 2005]. Copyright 2005 by the Optical Society of America.

of directly transmitted and reflected light. The highly focused beam also allows for localized excitation in a diffraction-limited spot area.

The excited SPPs will radiate back into the glass substrate in the form of leakage radiation at an angle $\theta_{\mathrm{SPP}} > \theta_c$, which can be collected through the immersion oil layer via the objective. Fig. 3.10 shows images of leakage radiation for SPP excited using a white-light continuum, tracing the path of the excited SPPs (in TM polarization only), since the intensity of the leakage radiation is proportional to the intensity of the SPPs themselves. This scheme is especially convenient for the excitation of a continuum of SPPs at different frequencies and the subsequent determination of their propagation lenghts.

3.5 Near-Field Excitation

Excitation schemes such as prism or grating coupling excite SPPs over a macroscopic area defined by the dimensions of the (at best diffraction-limited) spot of the coupling beam of wavelength λ_0. In contrast, near-field optical microscopy techniques allow for the *local* excitation of SPPs over an area $a << \lambda_0$, and can thus act as a point source for SPPs [Hecht et al., 1996]. Fig. 3.11 sketches the typical geometry: a small probe tip of aperture size $a \lesssim \lambda_{\mathrm{SPP}} \lesssim \lambda_0$ illuminates the surface of a metal film in the near field. Due to the small aperture size, the light ensuing from the tip will consist of wave vector components $k \gtrsim \beta \gtrsim k_0$, thus allowing phase-matched excitation of SPPs with propagation constant β. Due to the ease of lateral positioning of such probes in scanning near-field optical microscopes, SPPs at different locations of the metal surface can be excited.

A typical near-field optical setup suitable for local SPP excitation is shown in Fig. 3.12. SPPs propagating from the illumination spot can be conveniently imaged by collecting the leakage radiation into the substrate of refracting index n occurring at the SPP angle θ_{SPP} defined earlier. Called *forbidden light* by the authors of this study [Hecht et al., 1996] due to the fact that it is radiated outside

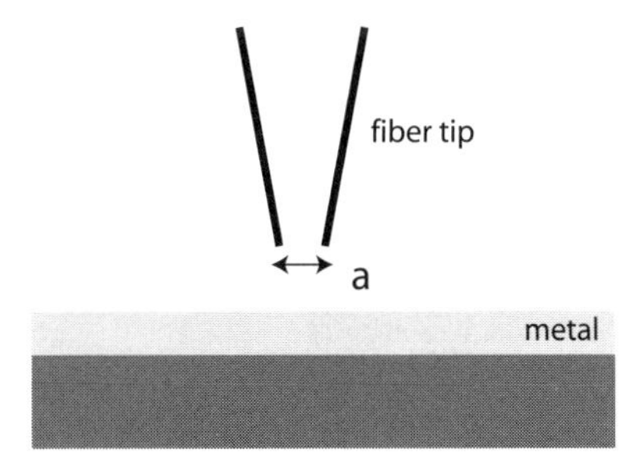

Figure 3.11. Local excitation of SPPs using near-field illumination with a sub-wavelength aperture.

the air light cone, this radiation can be either collected using a suitable mirror arrangement, or by using a collection objective with a high numerical aperture.

Fig. 3.13 shows two typical images of SPPs propagating away from the local illumination area. The two light jets emerging from the illumination spot are in the direction of the polarization of the electric field, due to the character of the SPPs as a mainly longitudinal electromagnetic surface wave for excitation close to ω_{sp}. The intensity variation of the SPPs can be fitted by

$$I_{SPP} \propto \frac{e^{-\rho/L}}{\rho} \cos^2 \phi, \tag{3.4}$$

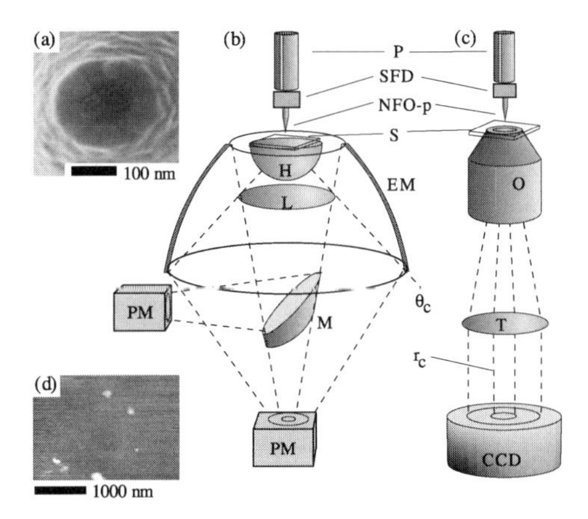

Figure 3.12. Near-field optical excitation of SPPs. (a) Scanning electron microscopy image of the aperture of a near-field fiber probe. (b) and (c) Two optical setups of the excitation of SPPs and the collection of light radiated into the substrate in the far field. (d) Topography of a silver film used as a sample (roughness 1 nm, height of protrusions 40 nm). More details about the setup can be found in [Hecht et al., 1996]. Reprinted with permission from [Hecht et al., 1996]. Copyright 1996 by the American Physical Society.

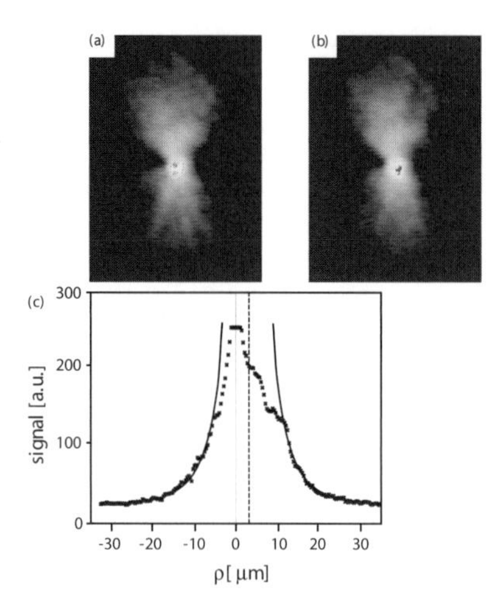

Figure 3.13. Spatial intensity distribution of SPPs on a silver film at $\lambda = 633$ nm. (a), (b) are 50μm $\times$ 70μm images collected in the far field corresponding to two different locations of the exciting near-field probe. (c) Cross section through the intensity profile along the main symmetry axes of the spots and analytical fit using (3.4). Reprinted with permission from [Hecht et al., 1996]. Copyright 1996 by the American Physical Society.

where ρ and ϕ are polar coordinates and L the intensity decay constant of the propagating SPP. As expected, the intensity distribution resembles that of damped radiation from a two-dimensional point dipole.

Using this local excitation scheme, the effect of surface roughness on the SPP propagation and the scattering at individual surface defects can be studied with high spatial resolution. Apart from the excitation of propagating SPPs, near-field illumination also allows for the excitation and spectral analysis of localized surface plasmon modes in individual metal nanostructures, which will be discussed in chapter 10.

3.6 Coupling Schemes Suitable for Integration with Conventional Photonic Elements

While the optical excitation schemes described above are suitable for the investigation of SPP propagation and functional plasmonic structures in proof-of-concept characterizations, practical applications of SPPs in integrated photonic circuits will require high-efficiency (and ideally high-bandwidth) coupling schemes. Preferably, the plasmonic components should allow efficient matching with conventional dielectric optical waveguides and fibers, which would in such a scenario be used to channel energy over large distances to plasmon waveguides and cavities. The latter will then enable high-confinement guiding and localized field-enhancement [Maier et al., 2001], for example for the routing of radiation to single molecules.

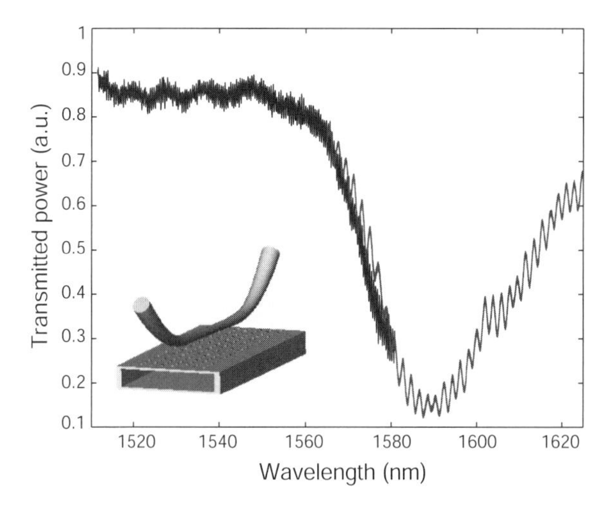

Figure 3.14. Excitation of SPPs propagating on a metal nanoparticle plasmon waveguide situated on a thin silicon membrane using a fiber taper (sketch in inset). The transmission spectrum shows the power transmitted through the taper past the coupling region, demonstrating a power transfer efficiency of 75% at $\lambda_0 = 1590$ nm due to phase-matching. Reprinted with permission from [Maier et al., 2005]. Copyright 2005, American Institute of Physics.

One such coupling scheme is end-fire coupling, in which a free-space optical beam is focused on the end-facet of the desired waveguide. Rather than relying on phase-matching, this scheme operates via matching the spatial field distribution of the waveguide as much as possible by adjusting the beam width. For SPPs propagating at a single interface, Stegeman and co-workers demonstrated coupling efficiencies up to 90% using this technique [Stegeman et al., 1983]. In contrast to prism coupling, end-fire excitation allows for the excitation of truly bound modes that do not radiate into the substrate. End-fire coupling is also particularly useful and efficient for exciting the long-ranging SPP mode propagating along thin metal films embedded in a symmetric dielectric host. Due to the delocalized nature of this mode (see chapters 2 and 7), spatial mode matching works especially well in this case. Naturally however, for geometries showing field-localization below the diffraction limit such as metal/insulator/metal waveguides with a deep sub-wavelength dielectric core, the overlap between the excitation beam and the coupled SPP mode is very small, leading to low excitation efficiencies.

For SPPs with larger confinement, a convenient interfacing scheme makes use of optical fiber tapers brought into the immediate vicinity of the waveguide to enable phase-matched power transfer via evanescent coupling [Maier et al., 2004]. Fig. 3.14 shows as an example the spectral dependence of the power transmitted past the coupling region between a fiber taper and a metal nanoparticle waveguide fabricated on top of a thin silicon membrane. The drop in detected power at the end of the fiber at $\lambda = 1590$ nm is due to power transfer to the plasmon waveguide, in this case with a coupling efficiency of about 75% [Maier et al., 2005]. More details about this particular fiber-accessible plasmon waveguide can be found in chapter 7.

Chapter 4

IMAGING SURFACE PLASMON POLARITON PROPAGATION

After the presentation of various approaches to optically launch surface plasmon polaritons, we move on to a description of ways to image the confined fields and their propagation along the interface. While the successful excitation of SPPs using optical techniques such as prism or grating coupling can be deduced from a decrease in the intensity of the reflected light beam (chapter 3), a direct visualization of the SPPs propagating away from the excitation region is highly desirable. This way, the propagation length L can be determined, influenced both by the amount of absorption inside the metal and leakage radiation (if present). Also, the amount of in-plane confinement can be assessed. An investigation of the out-of-plane confinement allows the determination of $\hat{z}$, the extent to which the evanescent fields penetrate inside the dielectric side of the interface. We have already mentioned the fundamental trade-off between propagation length and confinement, which is of tantamount importance in the design of plasmon waveguides (chapter 7).

This chapter discusses four prominent approaches for SPP imaging - near-field optical microscopy, imaging based on either fluorescence or leakage radiation detection, as well as the related observation of scattered light. Of these four techniques, only near-field optical microscopy provides the sub-wavelength resolution required for the accurate determination of the loss/confinement ratio for spatially highly localized SPPs excited near ω_{sp} or in appropriate multilayer structures. Cathodoluminescence imaging will be discussed in chapter 10 on localized plasmon spectroscopy.

4.1 Near-Field Microscopy

A powerful technique to investigate SPPs propagating at the interface of a thin metal film and air with sub-wavelength resolution is near-field optical microscopy in collection mode, also called *photon scanning tunneling mi-*

croscopy. The latter term highlights the conceptual similarity with the scanning tunneling microscope (STM). In both cases, a sharp tip is brought into the immediate vicinity of the surface under study (Fig. 4.1) using an appropriate feedback-loop technique. Whereas a STM measures the current (induced by an applied voltage) caused by electron tunneling between the surface and an atomically sharp metal tip, a photon scanning tunneling microscope (PSTM) collects photons by coupling the evanescent *near field* above the surface to propagating modes inside a tapered optical fiber. The near-field optical tip (also called the *probe*) is usually fabricated by pulling or etching an optical fiber taper, and is often metalized at the end in order to suppress the coupling of diffracted light fields. The resolution of this technique is limited by the size of the tip's aperture, which can reach dimensions of only 50 nm or even less using etching (or more recently also microfabrication) techniques. In addition to metal-coated probes, uncoated probes are also frequently used, which have a higher collection efficiency and have been shown to image different components of the electromagnetic field around nanostructures than probes coated with a conductive layer [Dereux et al., 2001].

In order to study the confinement and propagation of SPPs using this scheme, the tip has to be brought within a sufficiently close distance to the flat metal surface so that it is immersed in the evanescent tail of the SPP field, i.e. within a distance $\hat{z}$ (calculated using (2.12)). For studies of gold or silver films at visible frequencies, this requires a gap between the probe and the film on the order of 100 nm or less, which can be easily achieved using feedback techniques such as non-contact mode atomic force microscopy, shear or tuning force feedback, or by using the intensity of the collected light field itself as the feedback signal

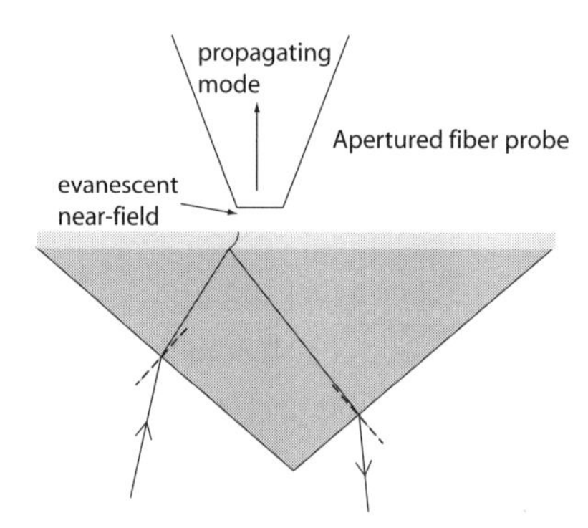

Figure 4.1. A typical setup for near-field optical imaging of SPP fields at a metal/air interface. The evanescent tail of the fields penetrating into the air is coupled to propagating modes in a tapered optical fiber tip. The SPPs can for example be excited via prism coupling (shown), a tightly focused beam, or particle impact.

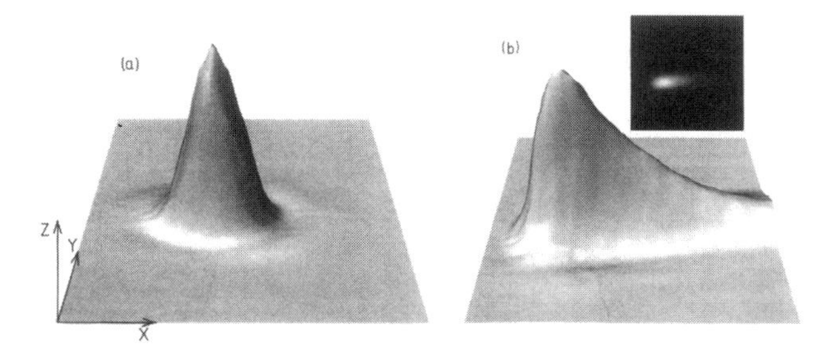

Figure 4.2. Near-field image of a HeNe laser beam ($\lambda = 633$ nm) internally incident on an uncoated (a) prism face and a prism face coated with a 53 nm thick silver film (b) at an angle greater than the critical angle (scan range 40x40 μm). The exponentially decaying tail in (b) is due to SPP propagation away from the excitation region. Reprinted with permission from [Dawson et al., 1994]. Copyright 1994 by the American Physical Society.

(akin to the STM, where the tunneling current proportional to the amount of collected electrons is used for this purpose).

In order not to interfere with the detection scheme, optical excitation of SPPs takes usually place via either prism coupling (Fig. 4.1) or tightly focused beams using an oil-immersion objective of high numerical aperture on the bottom side of the substrate. We note that the prism coupling scheme is not suitable for exciting SPPs of all possible propagation constants β, but only those within the window of leaky modes as discussed in chapter 3.

The very first studies of the physical properties of SPPs using near-field optical microscopy investigated the mode confinement at the interface of a thin silver film with air. SPPs were excited in prism coupling geometry, and the evanescent field on the air side probed via an apertured optical fiber tip. Without actually imaging the propagating fields using raster scanning, approaching and retracting the fiber probe confirmed the localization and corresponding enhancement of the electromagnetic field in the near-field region close to the surface [Marti et al., 1993]. Monitoring of the collected signal intensity at different heights above the surface allowed the determination of the penetration of the SPP fields into the air above the surface, confirming the spatial extent of the exponentially decaying field [Adam et al., 1993].

In addition to the investigation of out-of-plane confinement, the combination of near-field collection with raster scanning techniques enables the direct visualization of propagating SPPs. Dawson and co-workers used a PSTM to spatially image the propagation of SPPs excited using prism coupling on a thin silver film [Dawson et al., 1994]. Fig. 4.2b shows a three-dimensional rendering of the intensity collected in the near-field above the film surface. An excitation wavelength $\lambda_0 = 633$ nm in the visible regime ensured good confinement to the interface ($\hat{z} \approx 420$ nm calculated using (2.12)). As a control experiment,

Fig. 4.2a shows the evanescent field above a bare prism surface under the same excitation conditions. Clearly, for the silver-coated prism propagation of electromagnetic energy away from the excitation spot is visible. Experiments such as this enable the direct determination of the SPP propagation length L by fitting the exponential tail starting at the SPP launching point. In this case, the propagation length of the silver/air SPP was determined to be 13.2 μm, in good agreement with theoretical modeling. Also, the in-plane spread of the SPP as it propagates away from the excitation region can be monitored.

Collection-mode near-field optical microscopy has ever since these initial investigations been extensively employed for studies of SPP propagation, most prominently in a context of waveguiding along metal stripes, where the transverse extent of the SPP is limited by the stripe width (chapter 7). This has enabled the determination of the trade-off between propagation length and out-of-plane as well as lateral confinement, and additionally investigations of functional waveguide devices such as reflectors or Bragg mirrors. For example, near-field imaging allowed the direct visualization of interference patterns between co- and counterpropagating SPP waves. Some of these studies will be presented in chapter 7 on plasmon waveguides.

Near-field probing has also proved very useful for the assessment of scattering losses on structured metal surfaces [Bouhelier et al., 2001] as well as for the determination of the dispersion properties of SPPs at curved surfaces [Passian et al., 2004]. It has to be noted that the presence of the probing tip can influence the dispersion, but for dielectric tips this effect can often be neglected [Passian et al., 2005].

As might be expected, near-field optical microscopy is also often the method of choice for studies of localized surface plasmons in metal nanoparticles or ensembles of metal nanostructures (chapter 5). In these experiments, the light path is usually reversed: By not collecting but illuminating the metal structure under study via light emanating through the sub-wavelength aperture of a fiber tip, near-field optical spectroscopy of the localized modes is possible, in addition to imaging of the spatial field distribution. Examples will be presented in chapter 10 on spectroscopy and sensing.

In this *illumination mode*, the fiber probe effectively acts as a local dipolar source for the excitation of surface plasmons (or propagating SPPs as described in the previous chapter). Information about the electromagnetic structure of the surface can be extracted from the transmitted or reflected light collected using an objective in the far field. Apart from photon collection in the far field, the metal film structure under investigation can also be directly mounted on the photodiode itself, as shown by Dragnea and co-workers , which used this geometry for the study of SPP propagation in sub-wavelength slits on a flat metal film [Dragnea et al., 2003].

4.2 Fluorescence Imaging

Instead of locally collecting the optical near field of SPPs using the apertured fiber tip of a near-field optical microscope, emitters such as quantum dots or fluorescent molecules can be directly placed into the evanescent tail of the SPP field. If the frequency of the propagating SPPs lies within the broad spectral absorption band of the emitters, their excitation via SPPs is possible, and the intensity of the emitted fluorescence radiation is proportional to the intensity of the local field at the position of the emitters. Therefore, SPP propagation on a metal/air interface can be mapped by coating the surface with a dielectric film doped with emitters. If the layer is sufficiently thin and of low refractive index (e.g. quantum dots embedded in a polymer, or monolayers of fluorescent molecules), the alteration of the SPP dispersion due to the covering layer is small.

As will be discussed in more detail in chapter 9, fluorescent molecules placed into the near field of propagating SPPs (and also that of localized plasmons) show an enhancement of their fluorescence yield if care is taken to counteract non-radiative quenching. This can be achieved by inserting a thin spacer layer on the order of a few nanometers between the metal film sustaining the SPPs and the fluorescent molecules to inhibit non-radiative energy transfer.

Ditlbacher and co-workers used this concept for the imaging of SPPs excited on a 70 nm thin silver film by focusing a laser beam ($\lambda_0 = 514$ nm, $P =$

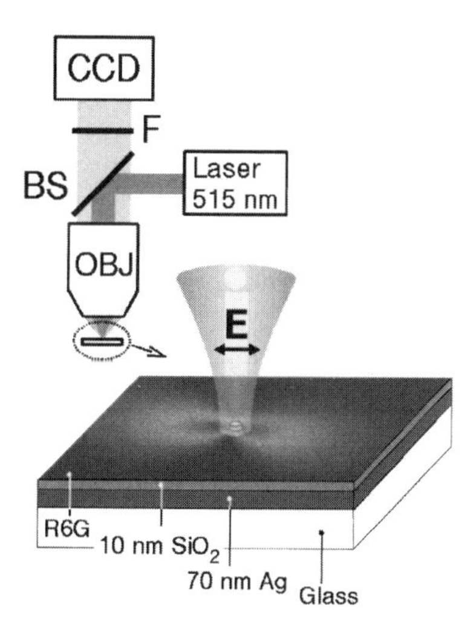

Figure 4.3. Fluorescence imaging of SPP fields. A SPP on a 70 nm silver film is excited via illumination of a nanoparticle (phase-matching via a defect) using a 100× objective, and the field distribution of the mode imaged by detecting the fluorescent emission of a coating layer doped with Rhodamin 6G. Reprinted with permission from [Ditlbacher et al., 2002a]. Copyright 2002, American Institute of Physics.

5 mW) on wire or nanoparticle surface defects created using electron beam lithography (Fig. 4.3) [Ditlbacher et al., 2002a]. The metal film was coated with a sub-monolayer of Rhodamine 6G molecules to enable the determination of the spatial structure of the SPP fields. In order to reduce quenching due to intermolecular interactions and non-radiative transitions to the metal film, the molecular density was chosen to be sufficiently small and a 10 nm thin SiO_2 spacer layer inserted between the molecular film and the silver substrate. CCD images of the fluorescence signal collected via a dichroic mirror are shown in Fig. 4.4. The intensity distribution correlates well with the pathways expected for SPPs excited via surface defects (compare with [Hecht et al., 1996] and Fig. 3.13 of chapter 3).

Using this scheme, information about the lateral spatial confinement, the propagation distance and interference effects can be extracted in analogy to the direct probing of the near field using an apertured probe, albeit with a resolution of at best the diffraction limit. However, the effect of bleaching in regions of high field intensity has to be carefully taken into account for quantitative analysis.

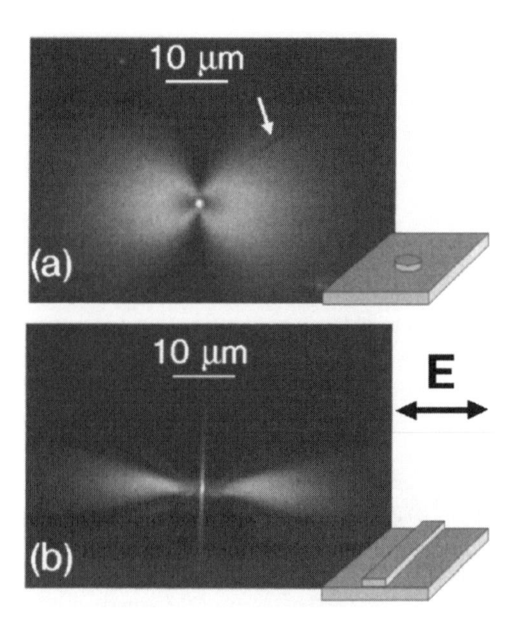

Figure 4.4. Fluorescence images of the intensity distribution of SPPs excited by illumination of (a) a silver nanoparticle (diameter 200 nm, height 60 nm), and (b) a silver nanowire (width 200 nm, height 60 nm, length 20 μm). The particles are situated on a continuous silver film supporting SPPs. Reprinted with permission from [Ditlbacher et al., 2002a]. Copyright 2002, American Institute of Physics.

4.3 Leakage Radiation

The dispersion curve of SPPs excited at the air interface of a metal film lies outside the light cone defined by $k = n_{\text{air}}\omega/c$, and the mode does not suffer radiation loss into the air region (for a perfectly flat interface neglecting roughness). However, energy can be lost into a supporting substrate of higher index n_s. This radiation loss occurs at all points of the dispersion curve that lie to the left of the light line of the substrate $k_s = n_s\omega/c$, as indicated in Figure 4.5. Therefore, for SPP excited in the region of propagation constants β defined by

$$k_0 < \beta < k_0 n_s, \tag{4.1}$$

leakage radiation into the substrate provides a second loss channel in addition to the inherent absorptive losses.

We have seen in the preceding chapter that leaky SPPs are inherently excited using prism coupling, and that the leakage radiation into the prism interferes with the directly reflected beam. As pointed out, a zero reflection ensues only under the condition of critical coupling (see (3.1)), when the absorptive losses exactly equal the radiative losses, and all power is absorbed in the metal film. This is only achieved for a critical thickness of the metal film.

Apart from monitoring the efficiency of prism coupling, leakage radiation collection can be used for investigating SPPs excited by other means, such as tightly focused beams or gratings, as long as the excited wave vectors β lie within the substrate light cone, fulfilling (4.1).

A typical setup for the collection of leakage radiation is shown in Fig. 4.6 [Ditlbacher et al., 2003]. In this study, the intensity of the leakage radiation

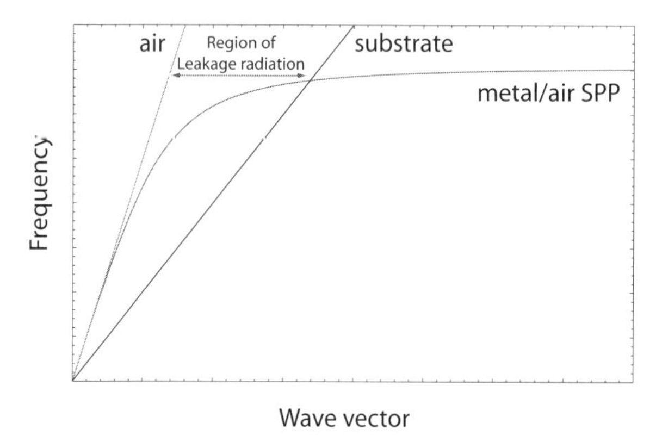

Figure 4.5. Generic dispersion relation of a SPP at a metal/air interface. In the region enclosed by the light lines of air and of the higher index substrate, the propagating SPPs lose energy via leakage radiation into the substrate light cone, which can be collected for imaging purposes.

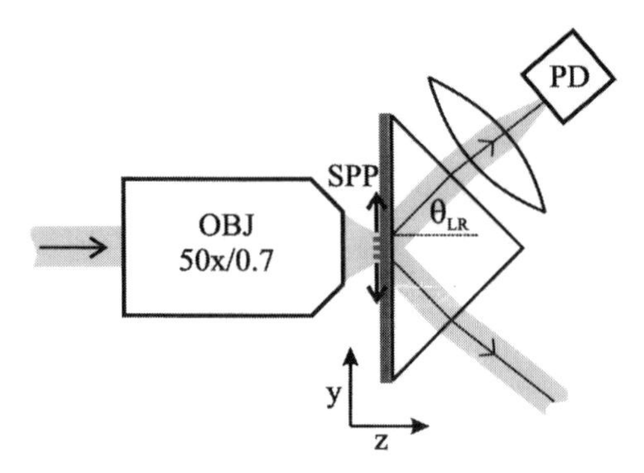

Figure 4.6. Experimental setup for leakage radiation imaging of SPP fields. Here, SPPs are excited using grating coupling, and the ensuing leakage radiation into the underlying prism collected using a photodiode. Reprinted with permission from [Ditlbacher et al., 2003]. Copyright 2003, American Institute of Physics.

was used to quantify the coupling efficiency of light to SPPs via a grating-like excitation scheme with a variable number of ridges spaced by a lattice constant Λ. We note that in this collection geometry, only half of the ensuing leakage radiation is collected via the underlying prism. With this technique, spatial intensity profiles can be obtained by varying the position of the sample with respect to the exciting laser beam. The amount of leakage radiation collected for films with one (a) and three coupling ridges of different lattice constants (b-d) is shown in Fig. 4.7. A maximum light-SPP coupling efficiency of 15% was achieved for a three-ridge sample of appropriate lattice constant. Naturally,

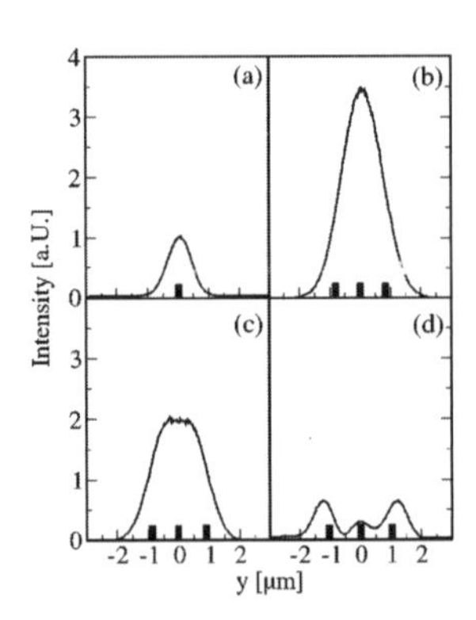

Figure 4.7. Quantifying coupling efficiency of a finite metal grating via collection of leakage radiation. The graphs show the experimentally observed distribution of leakage radiation vs. sample position (Fig. 4.6) for a single-ridge sample (a) and samples with three ridges of different lattice constants (b-d). The maximum intensity profile in (a) was normalized to 1. Reprinted with permission from [Ditlbacher et al., 2003]. Copyright 2003, American Institute of Physics.

the same setup can also be used to quantify the coupling efficiency of other methods, such as highly focused beams or coupling via inherent or designed surface roughness (chapter 3).

Leakage radiation also has to be considered in the design of plasmon waveguides. For example, all studies of laterally confined SPP propagation in metal stripes or nanowires where prism-coupling excitation has been employed inherently only investigate modes in the leaky region (4.1) described above. These *leaky waveguides* will be discussed in detail in chapter 7.

Apart from the observation of SPP propagation, leakage radiation imaging can also be used for the direct visualization of the SPP dispersion relation, which was demonstrated by Giannattasio and Barnes [Giannattasio and Barnes, 2005]. In this work, SPPs at the air interface of a 50 nm thick silver film were excited via a focused light beam using scattering from random surface roughness for phase-matching (Fig. 4.8). Leakage radiation into the silica substrate was directly imaged using a CCD camera glued to the underside of the substrate. For a flat film (Fig. 4.8a), the radiation is emitted within a cone of surface plasmon angle θ_{SPP} defined by $n_s k_0 \sin \theta_{\mathrm{SPP}} = \beta$, which intersects the plane of the CCD in a circular pattern. Light of different frequencies can be used for excitation, and the resulting wave vector β within the region (4.1) determined by the computation of the angle θ_{SPP} of leakage radiation from the

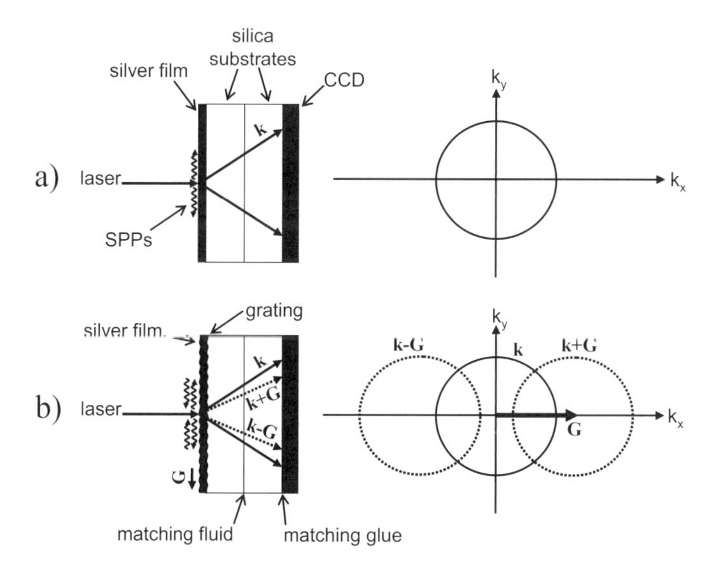

Figure 4.8. Experimental geometry of leakage radiation imaging for the determination of the SPP dispersion relation. (a) Planar silver surface: a single cone of light is emitted into the silica substrate. (b) Corrugated silver surface: the central cone is now intersected by other light cones due to SPPs scattered by the grating with Bragg vector **G**. Reproduced with permission from [Giannattasio and Barnes, 2005]. Copyright 2005, Optical Society of America.

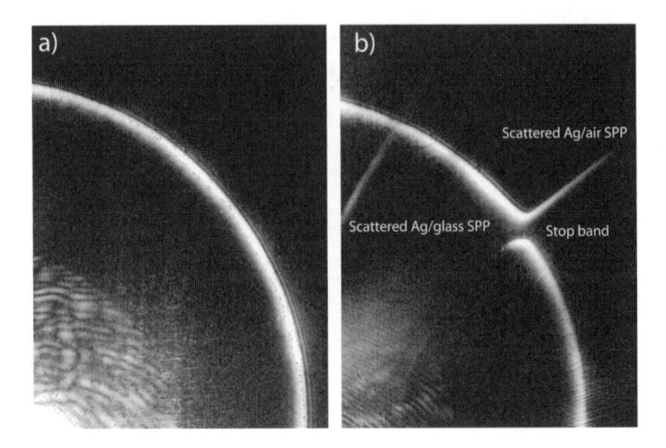

Figure 4.9. Direct image of the conical radiation in k-space sketched in Fig. 4.8 recorded by a CCD array. (a) Planar sample. (b) Imaging of the stop band emerging at the intersection between the two cones in k-space. Reproduced with permission from [Giannattasio and Barnes, 2005]. Copyright 2005, Optical Society of America.

radius of the imaged circle. Fig. 4.9a shows an image of parts of the circumference of the cone obtained by collecting the leakage radiation, confirming the usefulness of this method for the determination of the SPP dispersion relation.

This scheme allows a convenient way for the determination of the more complex dispersion relation of a structured metal surface. For a surface with regular, one-dimensional corrugations with grating constant a (corresponding to a reciprocal grating vector $\mathbf{G} = 2\pi/a$), perpendicular incidence of the exciting laser light leads to leakage radiation into a central light cone (corresponding to zero-order scattering) intersected by other cones ensuing from scattering SPPs with wave vectors $\mathbf{k} \pm \mathbf{G}$ (Fig. 4.8b). This leads to the formation of band gaps for SPP propagation at the intersection of adjacent cones, which are clearly visible in Fig. 4.9b as disruptions of the central circle. Additionally, scattering pathways both into the air and substrate layers are visible in these images, in the form of straight, jet-like lines.

4.4 Scattered Light Imaging

The propagation of SPPs at the air interface of metal films can often be simply imaged by collecting the light lost to radiation due to scattering at random (or indeed designed) surface protrusions. Scattering at these localized bumps allows SPPs with wave vector $\beta > k_0$ to acquire a momentum component Δk_x, which can lower β into the region within the air light cone (see equation (3.3)), leading to coupling to the radiation continuum and thus the emission of photons. For increasingly flat films with good surface quality, the amount of scattering is reduced, making a detailed determination of the properties of the SPPs such as their propagation length difficult.

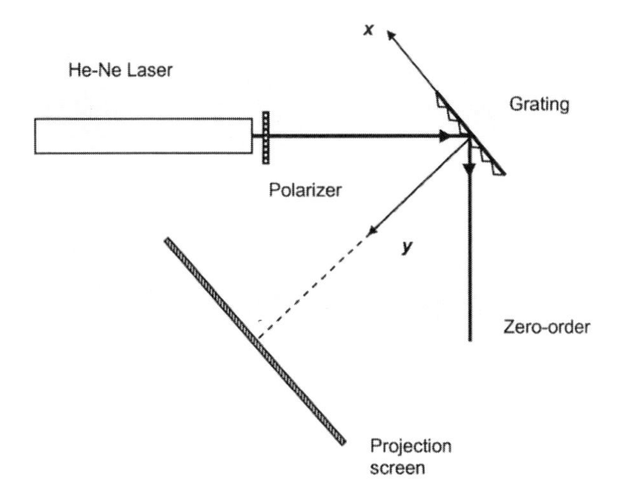

Figure 4.10. Experimental setup for the observation of the diffuse scattering background. Reprinted with permission from [Depine and Ledesma, 2004]. Copyright 2004, Optical Society of America.

The observation of light scattering from random roughness can also be used to map out the dispersion relation of SPPs on modulated surfaces. Depine and Ledesma used this method to determine the band gaps of SPPs for a metal surface corrugated with a blazed grating, by observing the so-called *diffuse light bands* [Depine and Ledesma, 2004]. These arise due to scattering from the random roughness of the grating. The experimental setup is very simple and shown in Fig. 4.10. A SPP is excited by focusing a laser beam under an angle θ to the surface normal onto the grating, and the scattered light is projected onto a screen parallel to the substrate.

It has been shown that a blazed grating leads to polarization conversion of the incoming and reflected light beam, mediated via SPPs, even when β is completely parallel to the grooves of the grating [Watts and Sambles, 1997]. A map of the reciprocal space (i.e., a two-dimensional plot of the in-plane components of β) is obtained by recording the intensity of the specular reflection versus incidence angle θ and the angle ϕ between β and the Bragg vector of the grating.

Depine and Ledesma have shown that the observation of the diffuse background does not necessitate such angular scanning in ϕ, which is now provided by scattering at the inherent surface roughness.

The obtained intensity maps of the in-plane components of β are presented in Fig. 4.11 both for light incident under TM (a) and TE (b) polarization. The observed structure corresponds well to a calculation of the reciprocal space of the electromagnetic modes sustained by this system, and to an experimental determination using angular scanning [Watts and Sambles, 1997]. In these

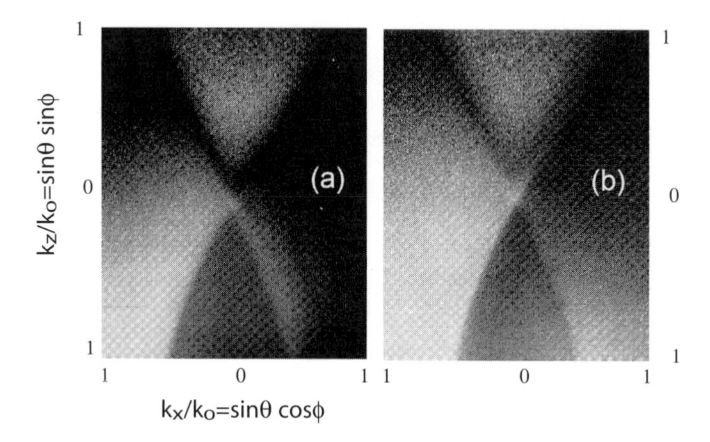

Figure 4.11. Reciprocal-space map of SPPs excited on a blazed grating for TM (a) and TE (b) polarized excitation beams. For details see text. Reprinted with permission from [Depine and Ledesma, 2004]. Copyright 2004, Optical Society of America.

pictures, the band gap can be determined by recording the minimum distance between the dark edges of forbidden β.

Chapter 5

LOCALIZED SURFACE PLASMONS

Here we introduce the second fundamental excitation of plasmonics - localized surface plasmons. We have seen in the preceding chapters that SPPs are propagating, dispersive electromagnetic waves coupled to the electron plasma of a conductor at a dielectric interface. Localized surface plasmons on the other hand are *non-propagating* excitations of the conduction electrons of metallic nanostructures coupled to the electromagnetic field. We will see that these modes arise naturally from the scattering problem of a small, sub-wavelength conductive nanoparticle in an oscillating electromagnetic field. The curved surface of the particle excerts an effective restoring force on the driven electrons, so that a resonance can arise, leading to field amplification both inside and in the near-field zone outside the particle. This resonance is called the *localized surface plasmon* or short *localized plasmon* resonance. Another consequence of the curved surface is that plasmon resonances can be excited by direct light illumination, in contrast to propagating SPPs, where the phase-matching techniques described in chapter 3 have to be employed.

We explore the physics of localized surface plasmons by first considering the interaction of metal nanoparticles with an electromagnetic wave in order to arrive at the resonance condition. Subsequent sections discuss damping processes, studies of plasmon resonances in particles of a variety of different shapes and sizes, and the effects of interactions between particles in ensembles. Other important nanostructures apart from solid particles that support localized plasmons are dielectric inclusions in metal bodies or surfaces, and nanoshells. The chapter closes with a brief look at the interaction of metal particles with gain media.

For gold and silver nanoparticles, the resonance falls into the visible region of the electromagnetic spectrum. A striking consequence of this are the bright colors exhibited by particles both in transmitted and reflected light, due to resonantly enhanced absorption and scattering. This effect has found applications for many hundreds of years, for example in the staining of glass for windows or ornamental cups. We will look at a number of more modern applications of localized plasmon resonances such as emission enhancement and optical sensing in chapters 9 and 10.

5.1 Normal Modes of Sub-Wavelength Metal Particles

The interaction of a particle of size d with the electromagnetic field can be analyzed using the simple *quasi-static approximation* provided that $d \ll \lambda$, i.e. the particle is much smaller than the wavelength of light in the surrounding medium. In this case, the phase of the harmonically oscillating electromagnetic field is practically constant over the particle volume, so that one can calculate the spatial field distribution by assuming the simplified problem of a particle in an electrostatic field. The harmonic time dependence can then be added to the solution once the field distributions are known. As we will show below, this lowest-order approximation of the full scattering problem describes the optical properties of nanoparticles of dimensions below 100 nm adequately for many purposes.

We start with the most convenient geometry for an analytical treatment: a homogeneous, isotropic sphere of radius a located at the origin in a uniform, static electric field $\mathbf{E} = E_0\hat{\mathbf{z}}$ (Fig. 5.1). The surrounding medium is isotropic and non-absorbing with dielectric constant ε_m, and the field lines are parallel to the z-direction at sufficient distance from the sphere. The dielectric response of the sphere is further described by the dielectric function $\varepsilon(\omega)$, which we take for the moment as a simple complex number ε.

In the electrostatic approach, we are interested in a solution of the *Laplace equation* for the potential, $\nabla^2\Phi = 0$, from which we will be able to calculate the electric field $\mathbf{E} = -\nabla\Phi$. Due to the azimuthal symmetry of the problem, the general solution is of the form [Jackson, 1999]

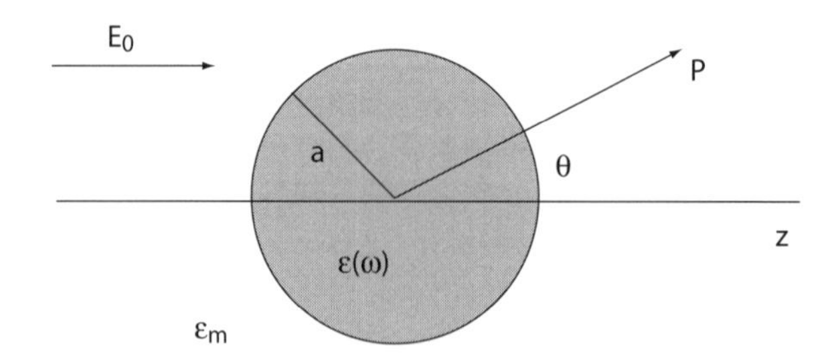

Figure 5.1. Sketch of a homogeneous sphere placed into an electrostatic field.

$$\Phi(r,\theta) = \sum_{l=0}^{\infty} \left[A_l r^l + B_l r^{-(l+1)} \right] P_l(\cos\theta), \tag{5.1}$$

where $P_l(\cos\theta)$ are the Legendre Polynomials of order l, and θ the angle between the position vector $\mathbf{r}$ at point P and the z-axis (Fig. 5.1). Due to the requirement that the potentials remain finite at the origin, the solution for the potentials Φ_{in} inside and Φ_{out} outside the sphere can be written as

$$\Phi_{\text{in}}(r,\theta) = \sum_{l=0}^{\infty} A_l r^l P_l(\cos\theta) \tag{5.2a}$$

$$\Phi_{\text{out}}(r,\theta) = \sum_{l=0}^{\infty} \left[B_l r^l + C_l r^{-(l+1)} \right] P_l(\cos\theta). \tag{5.2b}$$

The coefficients A_l, B_l and C_l can now be determined from the boundary conditions at $r \to \infty$ and at the sphere surface $r = a$. The requirement that $\Phi_{\text{out}} \to -E_0 z = -E_0 r \cos\theta$ as $r \to \infty$ demands that $B_1 = -E_0$ and $B_l = 0$ for $l \neq 1$. The remaining coefficients A_l and C_l are defined by the boundary conditions at $r = a$. Equality of the tangential components of the electric field demands that

$$-\frac{1}{a}\frac{\partial \Phi_{\text{in}}}{\partial \theta}\bigg|_{r=a} = -\frac{1}{a}\frac{\partial \Phi_{\text{out}}}{\partial \theta}\bigg|_{r=a}, \tag{5.3}$$

and the equality of the normal components of the displacement field

$$-\varepsilon_0 \varepsilon \frac{\partial \Phi_{\text{in}}}{\partial r}\bigg|_{r=a} = -\varepsilon_0 \varepsilon_{\text{m}} \frac{\partial \Phi_{\text{out}}}{\partial r}\bigg|_{r=a}. \tag{5.4}$$

Application of these boundary conditions leads to $A_l = C_l = 0$ for $l \neq 1$, and via the calculation of the remaining coefficients A_1 and C_1 the potentials evaluate to [Jackson, 1999]

$$\Phi_{\text{in}} = -\frac{3\varepsilon_m}{\varepsilon + 2\varepsilon_m} E_0 r \cos\theta \tag{5.5a}$$

$$\Phi_{\text{out}} = -E_0 r \cos\theta + \frac{\varepsilon - \varepsilon_m}{\varepsilon + 2\varepsilon_m} E_0 a^3 \frac{\cos\theta}{r^2}. \tag{5.5b}$$

It is interesting to interpret equation (5.5b) physically: Φ_{out} describes the superposition of the applied field and that of a dipole located at the particle center. We can rewrite Φ_{out} by introducing the dipole moment $\mathbf{p}$ as

$$\Phi_{\text{out}} = -E_0 r \cos\theta + \frac{\mathbf{p} \cdot \mathbf{r}}{4\pi\varepsilon_0\varepsilon_m r^3} \tag{5.6a}$$

$$\mathbf{p} = 4\pi\varepsilon_0\varepsilon_m a^3 \frac{\varepsilon - \varepsilon_m}{\varepsilon + 2\varepsilon_m}\mathbf{E}_0. \tag{5.6b}$$

We therefore see that the applied field induces a dipole moment inside the sphere of magnitude proportional to $|\mathbf{E}_0|$. If we introduce the polarizability α, defined via $\mathbf{p} = \varepsilon_0\varepsilon_m\alpha\mathbf{E}_0$, we arrive at

$$\alpha = 4\pi a^3 \frac{\varepsilon - \varepsilon_m}{\varepsilon + 2\varepsilon_m}. \tag{5.7}$$

Equation (5.7) is the central result of this section, the (complex) polarizability of a small sphere of sub-wavelength diameter in the electrostatic approximation. We note that it shows the same functional form as the Clausius-Mossotti relation [Jackson, 1999].

Fig. 5.2 shows the absolute value and phase of α with respect to frequency ω (in energy units) for a dielectric constant varying as $\varepsilon(\omega)$ of the Drude form (1.20), in this case fitted to the dielectric response of silver [Johnson and Christy, 1972]. It is apparent that the polarizability experiences a resonant enhancement under the condition that $|\varepsilon + 2\varepsilon_m|$ is a minimum, which for the case of small or slowly-varying $\text{Im}[\varepsilon]$ around the resonance simplifies to

$$\text{Re}[\varepsilon(\omega)] = -2\varepsilon_m. \tag{5.8}$$

This relationship is called the Fröhlich condition and the associated mode (in an oscillating field) the *dipole surface plasmon* of the metal nanoparticle. For a sphere consisting of a Drude metal with a dielectric function (1.20) located in air, the Fröhlich criterion is met at the frequency $\omega_0 = \omega_p/\sqrt{3}$. (5.8) further expresses the strong dependence of the resonance frequency on the dielectric

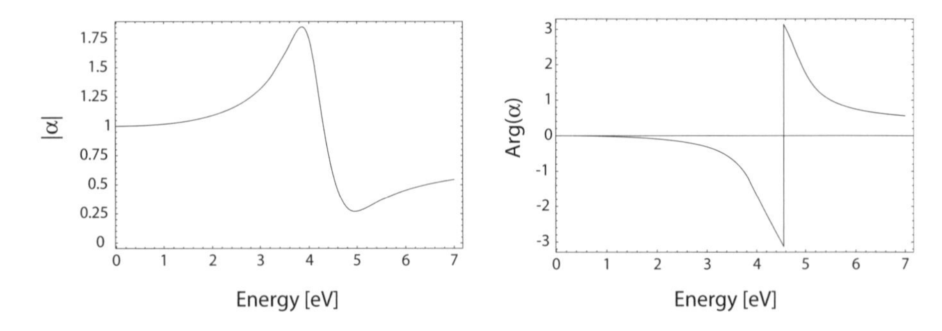

Figure 5.2. Absolute value and phase of the polarizability α (5.7) of a sub-wavelength metal nanoparticle with respect to the frequency of the driving field (expressed in eV units). Here, $\varepsilon(\omega)$ is taken as a Drude fit to the dielectric function of silver [Johnson and Christy, 1972].

environment: The resonance red-shifts as ε_m is increased. Metal nanoparticles are thus ideal platforms for optical sensing of changes in refractive index, which will be discussed in chapter 10.

We note that the magnitude of α at resonance is limited by the incomplete vanishing of its denominator, due to $\mathrm{Im}\,[\varepsilon(\omega)] \neq 0$. This will be elaborated in the last section of this chapter on nanoparticles in gain media.

The distribution of the electric field $\mathbf{E} = -\nabla\Phi$ can be evaluated from the potentials (5.5) to

$$\mathbf{E}_{\mathbf{in}} = \frac{3\varepsilon_m}{\varepsilon + 2\varepsilon_m}\mathbf{E}_0 \tag{5.9a}$$

$$\mathbf{E}_{\mathbf{out}} = \mathbf{E}_0 + \frac{3\mathbf{n}\,(\mathbf{n}\cdot\mathbf{p}) - \mathbf{p}}{4\pi\varepsilon_0\varepsilon_m}\frac{1}{r^3}. \tag{5.9b}$$

As expected, the resonance in α also implies a resonant enhancement of both the internal and dipolar fields. It is this field-enhancement at the plasmon resonance on which many of the prominent applications of metal nanoparticles in optical devices and sensors rely.

Up to this point, we have been on the firm ground of electrostatics, which we will now leave when turning our attention to the electromagnetic fields radiated by a small particle excited at its plasmon resonance. For a small sphere with $a \ll \lambda$, its representation as an ideal dipole is valid in the quasi-static regime, i.e. allowing for time-varying fields but neglecting spatial retardation effects over the particle volume. Under plane-wave illumination with $\mathbf{E}(\mathbf{r}, t) = \mathbf{E}_0 e^{-i\omega t}$, the fields induce an oscillating dipole moment $\mathbf{p}\,(t) = \varepsilon_0\varepsilon_m\alpha\mathbf{E}_0 e^{-i\omega t}$, with α given by the electrostatic result (5.7). The radiation of this dipole leads to *scattering* of the plane wave by the sphere, which can be represented as radiation by a point dipole.

It is useful to briefly review the basics of the electromagnetic fields associated with an oscillating electric dipole. The total fields $\mathbf{H}(t) = \mathbf{H}e^{-i\omega t}$ and $\mathbf{E}(t) = \mathbf{E}e^{-i\omega t}$ in the near, intermediate and radiation zones of a dipole can be written as [Jackson, 1999]

$$\mathbf{H} = \frac{ck^2}{4\pi}(\mathbf{n}\times\mathbf{p})\frac{e^{ikr}}{r}\left(1 - \frac{1}{ikr}\right) \tag{5.10a}$$

$$\mathbf{E} = \frac{1}{4\pi\varepsilon_0\varepsilon_m}\left\{k^2(\mathbf{n}\times\mathbf{p})\times\mathbf{n}\frac{e^{ikr}}{r} + \left[3\mathbf{n}\,(\mathbf{n}\cdot\mathbf{p}) - \mathbf{p}\right]\left(\frac{1}{r^3} - \frac{ik}{r^2}\right)e^{ikr}\right\}, \tag{5.10b}$$

with $k = 2\pi/\lambda$ and $\mathbf{n}$ the unit vector in the direction of the point P of interest. In the near zone ($kr \ll 1$), the electrostatic result (5.9b) for the electric field is recovered,

$$\mathbf{E} = \frac{3\mathbf{n}(\mathbf{n} \cdot \mathbf{p}) - \mathbf{p}}{4\pi\varepsilon_0\varepsilon_m} \frac{1}{r^3} \tag{5.11a}$$

and the accompanying magnetic field present for oscillating fields amounts to

$$\mathbf{H} = \frac{i\omega}{4\pi}(\mathbf{n} \times \mathbf{p}) \frac{1}{r^2}. \tag{5.11b}$$

We can see that within the near field, the fields are predominantly electric in nature, since the magnitude of the magnetic field is about a factor $\sqrt{\varepsilon_0/\mu_0}\,(kr)$ smaller than that of the electric field. For static fields ($kr \to 0$), the magnetic field vanishes.

In the opposite limit of the radiation zone, defined by $kr \gg 1$, the dipole fields are of the well-known spherical-wave form

$$\mathbf{H} = \frac{ck^2}{4\pi}(\mathbf{n} \times \mathbf{p}) \frac{e^{ikr}}{r} \tag{5.12a}$$

$$\mathbf{E} = \sqrt{\frac{\mu_0}{\varepsilon_0\varepsilon_m}}\mathbf{H} \times \mathbf{n}. \tag{5.12b}$$

We will now leave this short summary of the properties of dipolar radiation, and refer to standard textbooks on electromagnetism such as [Jackson, 1999] for further particulars. From the viewpoint of optics, it is much more interesting to note that another consequence of the resonantly enhanced polarization α is a concomitant enhancement in the efficiency with which a metal nanoparticle scatters and absorbs light. The corresponding cross sections for scattering and absorption C_{sca} and C_{abs} can be calculated via the Poynting-vector determined from (5.10) [Bohren and Huffman, 1983] to

$$C_{\mathrm{sca}} = \frac{k^4}{6\pi}|\alpha|^2 = \frac{8\pi}{3}k^4a^6\left|\frac{\varepsilon - \varepsilon_m}{\varepsilon + 2\varepsilon_m}\right|^2 \tag{5.13a}$$

$$C_{\mathrm{abs}} = k\mathrm{Im}[\alpha] = 4\pi ka^3\mathrm{Im}\left[\frac{\varepsilon - \varepsilon_m}{\varepsilon + 2\varepsilon_m}\right]. \tag{5.13b}$$

For small particles with $a \ll \lambda$, the efficiency of absorption, scaling with a^3, dominates over the scattering efficiency, which scales with a^6. We point out that no explicit assumptions were made in our derivations so far that the sphere is indeed metallic. The expressions for the cross sections (5.13) are thus valid also for dielectric scatterers, and demonstrate a very important problem for practical purposes. Due to the rapid scaling of $C_{\mathrm{sca}} \propto a^6$, it is very difficult to pick out small objects from a background of larger scatterers. Imaging of nanoparticles with dimensions below 40 nm immersed in a background of larger scatterers can thus usually only be achieved using photothermal techniques relying on the slower scaling of the absorption cross section with size

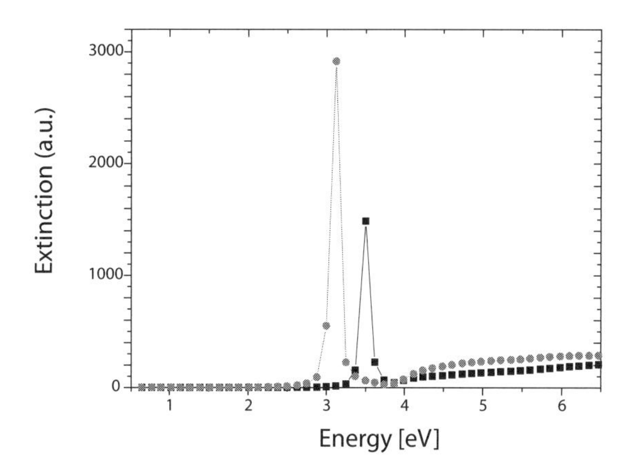

Figure 5.3. Extinction cross section calculated using (5.14) for a silver sphere in air (black curve) and silica (gray curve), with the dielectric data taken from [Johnson and Christy, 1972].

[Boyer et al., 2002], which will be elaborated on in chapter 10. Equations (5.13) also shows that indeed for metal nanoparticles both absorption and scattering (and thus extinction) are resonantly enhanced at the dipole particle plasmon resonance, i.e. when the Frölich condition (5.8) is met [Kreibig and Vollmer, 1995]. For a sphere of volume V and dielectric function $\varepsilon = \varepsilon_1 + i\varepsilon_2$ in the quasi-static limit, the explicit expression for the extinction cross section $C_{\text{ext}} = C_{\text{abs}} + C_{\text{sca}}$ is

$$C_{\text{ext}} = 9\frac{\omega}{c}\varepsilon_m^{3/2} V \frac{\varepsilon_2}{[\varepsilon_1 + 2\varepsilon_m]^2 + \varepsilon_2^2}. \tag{5.14}$$

Fig. 5.3 shows the extinction cross section of a silver sphere in the quasi-static approximation calculated using this formula for immersion in two different media.

We now relax the assumption of a spherical nanoparticle shape. However, it has to be pointed out that the basic physics of the localized surface plasmon resonance of a sub-wavelength metallic nanostructure is well described by this special case. A slightly more general geometry amenable to analytical treatment in the electrostatic approximation is that of an ellipsoid with semiaxes $a_1 \leq a_2 \leq a_3$, specified by $\frac{x^2}{a_1^2} + \frac{y^2}{a_2^2} + \frac{z^2}{a_3^2} = 1$. A treatment of the scattering problem in ellipsoidal coordinates [Bohren and Huffman, 1983] leads to the following expression for the polarizabilities α_i along the principal axes $(i = 1, 2, 3)$:

$$\alpha_i = 4\pi a_1 a_2 a_3 \frac{\varepsilon(\omega) - \varepsilon_m}{3\varepsilon_m + 3L_i(\varepsilon(\omega) - \varepsilon_m)} \tag{5.15}$$

L_i is a geometrical factor given by

$$L_i = \frac{a_1 a_2 a_3}{2} \int_0^\infty \frac{dq}{\left(a_i^2 + q\right) f(q)}, \tag{5.16}$$

where $f(q) = \sqrt{\left(q + a_1^2\right)\left(q + a_2^2\right)\left(q + a_3^2\right)}$. The geometrical factors satisfy $\sum L_i = 1$, and for a sphere $L_1 = L_2 = L_3 = \frac{1}{3}$. As an alternative, the polarizability of ellipsoids is also often expressed in terms of the *depolarization factors* $\tilde{L}_i$, defined via $E_{1i} = E_{0i} - \tilde{L}_i P_{1i}$, where E_{1i} and P_{1i} are the electric field and polarization induced inside the particle by the applied field E_{0i} along a principal axis i, respectively. $\tilde{L}$ is linked to L via

$$\tilde{L}_i = \frac{\varepsilon - \varepsilon_m}{\varepsilon - 1} \frac{L_i}{\varepsilon_0 \varepsilon_m}. \tag{5.17}$$

An important special class of ellipsoids are *spheroids*. For *prolate* spheroids, the two minor axes are equal ($a_2 = a_3$), while for *oblate* spheroids, the two major axes are of same size ($a_1 = a_2$). An examination of (5.15) reveals that a spheroidal metal nanoparticle exhibits two spectrally separated plasmon resonances, corresponding to oscillations of its conduction electrons along the major or minor axis, respectively. The resonance due to oscillations along the major axis can show a significant spectral red-shift compared to the plasmon resonance of a sphere of the same volume. Thus, plasmon resonances can be lowered in frequency into the near-infrared region of the spectrum using metallic nanoparticles with large aspect ratio. For a quantitative treatment, we note however that (5.15) is only strictly valid as long as the major axis is significantly smaller than the excitation wavelength.

Using a similar analysis, the problem of spheres or ellipsoids coated with a concentric layer of a different material can be addressed. Since core/shell particles consisting of a dielectric core and a thin, concentric metallic shell have recently attracted a great amount of interest in plasmonics due to the wide tunability of the plasmon resonance, we want to state the result for the polarizability of a coated sub-wavelength sphere with inner radius a_1, material $\varepsilon_1(\omega)$ and outer radius a_2, material $\varepsilon_2(\omega)$ [Bohren and Huffman, 1983]. The polarizability evaluates to

$$\alpha = 4\pi a_2^3 \frac{(\varepsilon_2 - \varepsilon_m)(\varepsilon_1 + 2\varepsilon_2) + f(\varepsilon_1 - \varepsilon_2)(\varepsilon_m + 2\varepsilon_2)}{(\varepsilon_2 + 2\varepsilon_m)(\varepsilon_1 + 2\varepsilon_m) + f(2\varepsilon_2 - 2\varepsilon_m)(\varepsilon_1 - \varepsilon_2)}, \tag{5.18}$$

with $f = a_1^3 / a_2^3$ being the fraction of the total particle volume occupied by the inner sphere.

5.2 Mie Theory

We have seen that the theory of scattering and absorption of radiation by a small sphere predicts a resonant field enhancement due to a resonance of the polarizability α (5.7) if the Frölich condition (5.8) is satisfied. Under these circumstances, the nanoparticle acts as an electric dipole, resonantly absorbing and scattering electromagnetic fields. This theory of the *dipole* particle plasmon resonance is strictly valid only for vanishingly small particles; however, in practice the calculations outlined above provide a reasonably good approximation for spherical or ellipsoidal particles with dimensions below 100 nm illuminated with visible or near-infrared radiation.

However, for particles of larger dimensions, where the quasi-static approximation is not justified due to significant phase-changes of the driving field over the particle volume, a rigorous *electrodynamic* approach is required. In a seminal paper, Mie in 1908 developed a complete theory of the scattering and absorption of electromagnetic radiation by a sphere, in order to understand the colors of colloidal gold particles in solution [Mie, 1908]. The approach of what is now know as *Mie theory* is to expand the internal and scattered fields into a set of *normal modes* described by vector harmonics. The quasi-static results valid for sub-wavelength spheres are then recovered by a power series expansion of the absorption and scattering coefficients and retaining only the first term.

Since Mie theory is treated in a variety of books such as [Bohren and Huffman, 1983, Kreibig and Vollmer, 1995] and a detailed knowledge of the higher order terms is not required for our purpose, we will not present it in this treatment, but rather examine the physical consequences of the first-order corrections to the quasi-static approximation.

5.3 Beyond the Quasi-Static Approximation and Plasmon Lifetime

Having obtained the general expressions (5.7) and (5.15) for the polarizability of a metal sphere and an ellipsoid in the quasi-static approximation, we will now analyze changes to the spectral position and width of the plasmon resonance with particle size not captured by this theory. Two regimes will be considered: Firstly, that of larger particles where the quasi-static approximation breaks down due to retardation effects, and secondly the regime of very small metal particles of radius $a < 10$ nm, where the particle dimensions are appreciably smaller than the mean free path of its oscillating electrons.

Starting with larger particles, a straight-forward expansion of the first TM mode of Mie theory yields for the polarizability of a sphere of volume V the expression [Meier and Wokaun, 1983, Kuwata et al., 2003]

$$\alpha_{\text{Sphere}} = \frac{1 - \left(\frac{1}{10}\right)(\varepsilon + \varepsilon_m)\, x^2 + O\left(x^4\right)}{\left(\frac{1}{3} + \frac{\varepsilon_m}{\varepsilon - \varepsilon_m}\right) - \frac{1}{30}(\varepsilon + 10\varepsilon_m)\, x^2 - i\frac{4\pi^2 \varepsilon_m^{3/2}}{3}\frac{V}{\lambda_0^3} + O\left(x^4\right)} V, \quad (5.19)$$

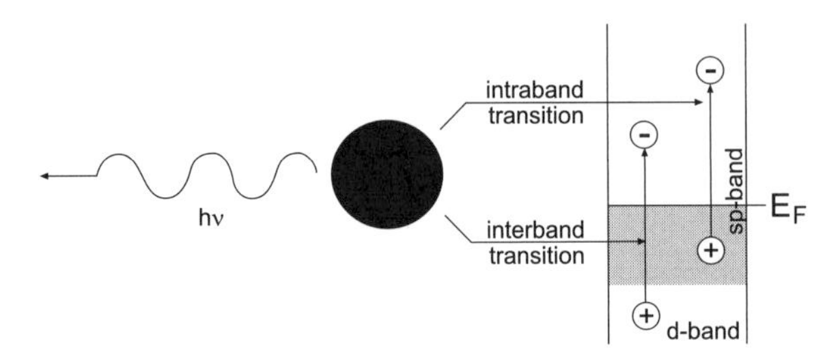

Figure 5.4. Schematic of radiative (left) and non-radiative (right) decay of particle plasmons.

where $x = \frac{\pi a}{\lambda_0}$ is the so called *size parameter*, relating the radius to the free-space wavelength. Compared to the simple quasi-static solution (5.7), a number of additional terms appear in the numerator and denominator of (5.19), each having a distinct physical significance. The term quadratic in x in the numerator includes the effect of retardation of the exciting field over the volume of the sphere, leading to a shift in the plasmon resonance. The quadratic term in the denominator also causes an energy shift of the resonance, due to the retardation of the *depolarization field* [Meier and Wokaun, 1983] inside the particle. For Drude and the noble metals, the overall shift is towards lower energies: the spectral position of the dipole resonance red-shifts with increasing particle size. Intuitively, this can be understood by recognizing that the distance between the charges at opposite interfaces of the particle increases with its size, thus leading to a smaller restoring force and therefore a lowering of the resonance frequency. This red-shift also implies that effects of interband transitions (described by an increase in $\mathrm{Im}\,[\varepsilon_2]$) not captured by the Drude theory decrease as the plasmon resonance moves away from the interband transition edge.

The quadratic term in the denominator also increases the magnitude of the polarization, and thus inherently lessens the influence of the absorption due to the imaginary part of ε. However, this increase in strength is counteracted by the third, completely imaginary term in the denominator, which accounts for *radiation damping*. An inclusion of terms of higher order in expression (5.19) will lead to the occurance of higher-order resonances, which will be touched upon in the next section.

Radiation damping is caused by a direct radiative decay route of the coherent electron oscillation into photons [Kokkinakis and Alexopoulos, 1972], and is the main cause of the weakening of the strength of the dipole plasmon resonance as the particle volume increases [Wokaun et al., 1982]. Thus, despite the fact that an increase in particle volume decreases the strength of the non-radiative decay pathway (namely absorption), a significant broadening of the plasmon resonance sets in.

We can summarize that the plasmon resonance of particles beyond the quasi-static regime is damped by two competing processes (Fig. 5.4): a radiative decay process into photons, dominating for larger particles, and a non-radiative process due to absorption. The non-radiative decay is due to the creation of electron-hole pairs via either intraband excitations within the conduction band or interband transitions from lower-lying d-bands to the sp conduction band (for noble metal particles). More details on the physics of the damping can be found in [Link and El-Sayed, 2000, Sönnichsen et al., 2002b].

In order to arrive at a quantitative description, these two damping processes can be incorporated into a simple two-level model of the plasmon resonance, developed by Heilweil and Hochstrasser [Heilweil and Hochstrasser, 1985]. Using it, the homogeneous linewidth Γ of the plasmon resonance, which can be determined using for example extinction spectroscopy, can be related to the internal damping processes via the introduction of a *dephasing time* T_2. In energy units, the relation between Γ and T_2 is

$$\Gamma = \frac{2\hbar}{T_2}. \tag{5.20}$$

We note that in analogy to dielectric resonators, the strength of a plasmon resonance can also be expressed using the notion of a *quality factor* Q, given by $Q = E_{\mathrm{res}}/\Gamma$, where E_{res} is the resonant energy.

In this theory, dephasing of the coherent excitation is either due to energy decay, or scattering events that do not change the electron energy but its momentum. This can be expressed by relating T_2 to a population relaxation or decay time T_1, describing both radiative and non-radiative energy loss processes, and a *pure dephasing time* T_2^* resulting from elastic collisions:

$$\frac{1}{T_2} = \frac{1}{2T_1} + \frac{1}{T_2^*}. \tag{5.21}$$

Via an examination of the details of plasmon decay, for example with pump-probe experiments [Link and El-Sayed, 2000], it can be shown that in general $T_2^* \gg T_1$ [Link and El-Sayed, 2000], so that $T_2 = 2T_1$. For small gold and silver nanoparticles, in general 5 fs $\leq T_2 \leq$ 10 fs, depending on size and the surrounding host material. Fig. 5.5 shows observed dephasing times for gold and silver nanospheres of varying diameter investigated using dark-field microscopy. In this figure, the magnitude of the plasmon decay is plotted in terms of Γ and T_2, related via (5.20). As apparent, in the case of gold the observed decay times can be well explained using Mie theory and the measured dielectric data [Johnson and Christy, 1972]. In the case of silver however, the agreement is less good, and especially for small silver spheres a significant decrease in dephasing time is observed, possibly due to damping processes at the particle surface.

The relative contributions of radiative and non-radiative pathways to the decay time T_1 is of importance for applications where sample heating or quench-

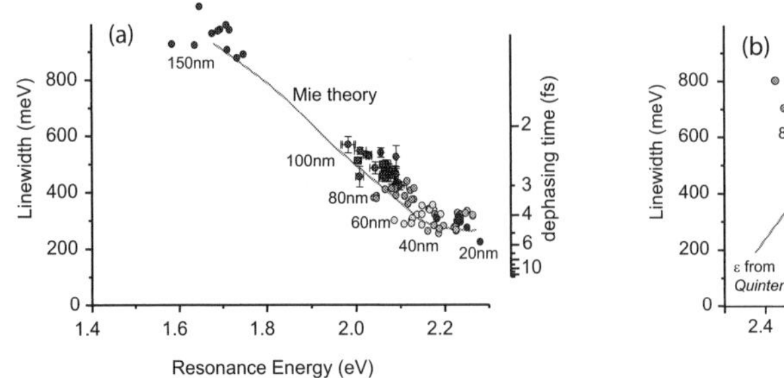

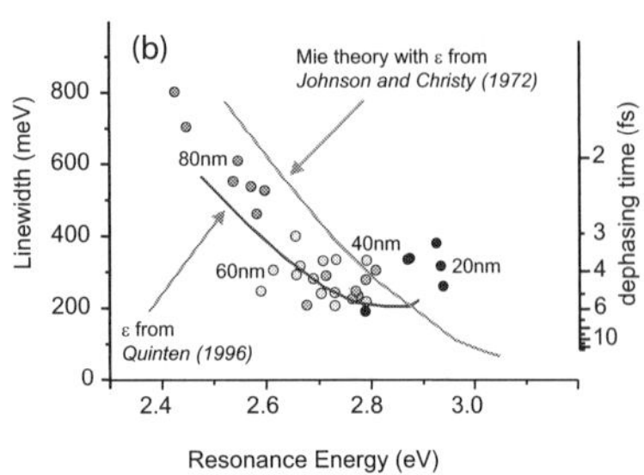

Figure 5.5. Linewidth of plasmon resonances of gold (a) and silver (b) nanospheres measured using dark-field microscopy, compared with predictions from Mie theory [Sönnichsen et al., 2002a]. Copyright 2002, Institute of Physics.

ing of fluorescence of molecules in the vicinity of the metal nanostructures are to be avoided. In this case, the radiative decay pathway should dominate. In order to achieve this, Sönnichsen and co-workers performed a study aimed at maximizing the radiative contribution $T_{1,r}$ to the total decay time over the non-radiative contribution $T_{1,nr}$ in gold nanorods of different aspect ratios [Sönnichsen et al., 2002b]. This corresponds to the maximization of the quantum efficiency η for resonant light scattering, given by

$$\eta = \frac{T_{1,r}^{-1}}{T_1^{-1}} = \frac{T_{1,r}^{-1}}{T_{1,r}^{-1} + T_{1,nr}^{-1}}. \tag{5.22}$$

In this study, the decay time of nanorods approached a limiting value $T_2 \approx$ 18 fs for a rod aspect ratio of 3:1, which is significantly larger than the dephasing time of gold nanospheres of similar volume (see Fig. 5.5). This is mainly due to a decrease in non-radiative damping caused by the change from the spherical to the spheroidal geometry: the long-axis mode shifts towards lower energies, thus limiting the influence of interband transitions.

We now turn the attention to the regime of very small metallic particles. For gold and silver particles of radius $a < 10$ nm, an additional damping process, loosely termed *chemical interface damping*, must be considered. Here, the rate of dephasing of the coherent oscillation is increased due to elastic scattering at the particle surface, since the size of the particle is substantially smaller than the electron mean free path (of the order of 30-50 nm). This could explain the observed decrease in decay time for small silver particles presented in Fig. 5.5. Empirically, the associated broadening of the experimentally observed plasmon linewidth Γ_{obs} can be modeled via [Kreibig and Vollmer, 1995]

$$\Gamma_{\text{obs}}(R) = \Gamma_0 + \frac{Av_F}{R}. \tag{5.23}$$

Here, Γ_0 describes the plasmon linewidth of particles that are outside the regimes where interface damping or radiation damping dominate, i.e. where Γ is defined by $\text{Im}\,[\varepsilon(\omega)]$ alone. v_F the Fermi velocity of the electrons, and $A \approx 1$ a factor incorporating details of the scattering process [Hövel et al., 1993]. In addition to the broadening of the resonance, shifts in resonance energy have also been reported for particles of dimensions below 10 nm. However, the direction of this shift seems to depend strongly on the chemical termination of the particle surface, and both blue- and red-shifts have been experimentally observed (for an overview see [Kreibig and Vollmer, 1995]).

While up to now our treatment of the interaction of a small metal particle with an incident electromagnetic wave has been purely classical, for particles with a radius of the order of or below 1 nm, quantum effects begin to set in. The reason that the quantized nature of the energy levels can be discarded down to this size scale is the large concentration of conduction electrons $n \approx 10^{23}$ cm^{-3} in metals. However, for small absolute numbers of electrons $N_e = nV$, the amount of energy gained by individual electrons per incident photon excitation, $\Delta E \approx \frac{\hbar\omega}{N_e}$, becomes significant compared to $k_B T$. In this regime the notion of a plasmon as a coherent electron oscillation breaks down, and the problem has to be treated using the quantum mechanical picture of a multiple-particle excitation. A description of these processes [Kreibig and Vollmer, 1995] lies outside the scope of this book.

5.4 Real Particles: Observations of Particle Plasmons

Localized plasmon resonances can readily be observed using far-field extinction microscopy on colloidal or nanofabricated metal nanostructures under illumination with visible light. A convenient way to create particles with a variety of shapes, albeit of an inherently planar nature, is electron beam lithography followed by a metal lift-off process. If far-field extinction microscopy is employed, the small size of nanoparticles with $d \ll \lambda_0$ compared to the at-best diffraction-limited illumination spot requires excitation of plasmons in arrays of particles of equal shape in order to achieve an acceptable signal-to-noise ratio in the extinction spectra. Typically, the particles are arranged on a square grid [Craighead and Niklasson, 1984], with a sufficiently large interparticle spacing to prevent interactions via dipolar coupling, which will be discussed in the next section. Despite the fact that the attenuation of the excitation beam is caused by absorption (and to a lesser degree scattering as long as $a \ll \lambda_0$) by multiple particles, the high reproducibility of particle shapes offered by electron beam lithography enables observations of resonance lineshapes approaching that of the homogeneous lineshape of a single particle.

Fig. 5.6 shows an example of extinction spectra of gold nanowires of various lengths fabricated using electron beam lithography and arranged in grids as described above. Since the nanowire length d is comparable or greater than λ_0, several resonances due to the excitation of higher-order oscillation modes are clearly visible. Due to the retardation effects outlined in the preceding section, the dipole resonance has experienced a profound red-shift to energies lower than those covered by the spectral range of the illumination source.

In contrast to far-field extinction microscopy techniques, far-field dark-field optical microscopy and near-field optical extinction microscopy enable the observation of plasmon resonances of a *single* particle. In dark-field optical microscopy, only the light scattered by the structure under study is collected in the detection path, while the directly transmitted light is blocked using a dark-field condenser. This enables the study of single particles dilutely dispersed on a substrate. Fig. 5.7 shows as an example the dipolar plasmon lineshapes of colloidal silver particles of different shapes. Other studies have investigated resonances in metal nanowires composed of segments of different metals [Mock et al., 2002b] and the influence of the refractive index on the plasmon resonance [Mock et al., 2003]. This scheme is particularly useful for biological sensing purposes, where resonance shifts due to binding events on single particles are monitored, which will be discussed in more detail in chapter 10.

In near-field optical spectroscopy, a thin (metalized or uncoated) fiber tip with an aperture on the order of 100 nm is brought into close proximity of the particle using an appropriate feedback scheme. The plasmon resonances can then be mapped out using either illumination through the tip and collection in the far-field, or evanescent illumination from the substrate side and light collection via the tip. For example, such investigations have enabled the determination of both the homogeneous linewidth Γ of a single nanoparticle [Klar

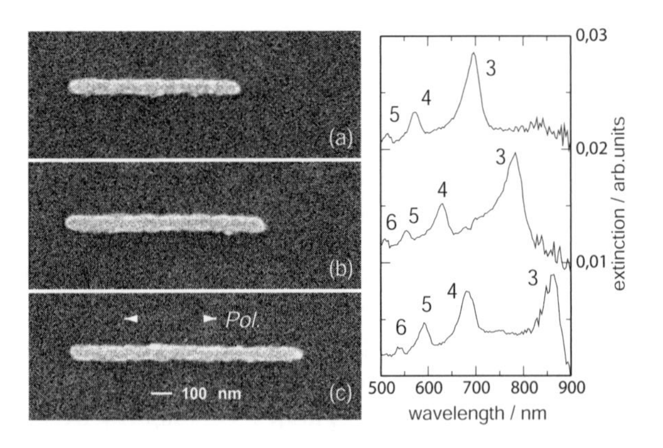

Figure 5.6. SEM images (left) and corresponding spectra (right) of gold nanowires excited with light polarized along their long axis of 790 nm (a), 940 nm (b), and 1090 nm (c). The length of the short axis and the height are 85 nm and 25 nm, respectively. Numbers at the spectral peaks indicate the order of the multipolar excitation. Reprinted with permission from [Krenn et al., 2000]. Copyright 2000, American Institute of Physics.

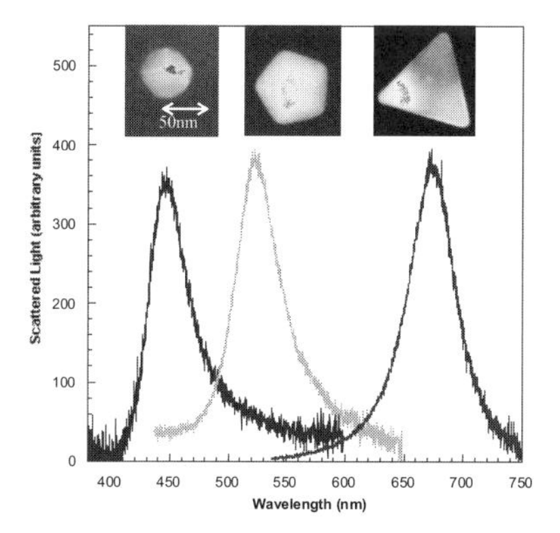

Figure 5.7. Scattering spectra of single silver nanoparticles of different shapes obtained in dark-field configuration. Reprinted with permission from [Mock et al., 2002a]. Copyright 2002, American Institute of Physics.

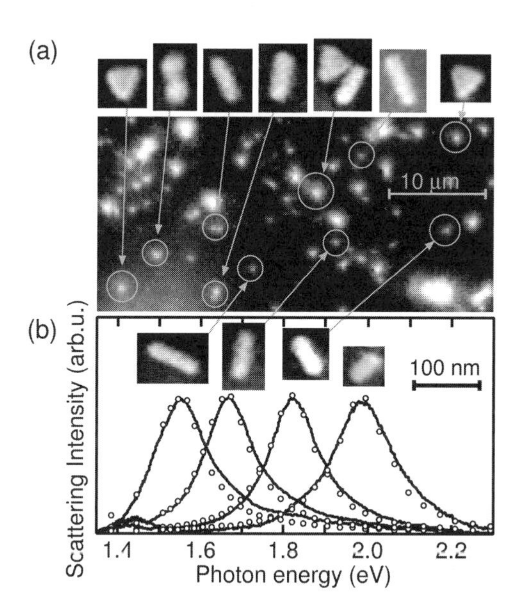

Figure 5.8. Optical dark field images together with SEM images of individual gold nanoparticles (a) and corresponding scattering spectra (b) for an incident light polarization along the long particle axis. Lines are experimental data, and circles cross sections calculated using the empirical formula (5.24). Reprinted with permission from [Kuwata et al., 2003]. Copyright 2003, American Institute of Physics.

et al., 1998] and the direct imaging of multipolar fields [Hohenau et al., 2005a], as well as the dispersion relation of gold nanorods [Imura et al., 2005]. More details of typical setups can be found in chapter 10 on spectroscopy.

We want to finish this section by presenting results from a comprehensive study of the influence of aspect ratio on the dipolar plasmon resonance in nanorods performed using dark-field optical spectroscopy [Kuwata et al., 2003]. Fig. 5.8 shows scattered light images and plasmon lineshapes (solid lines) of a variety of gold nanoparticles. Using this data, Kuwata and coworkers established an empirical extension of the formula for the polarizability for spherical particles (5.19) to ellipsoidal structures. For particles with volume V and size parameter x, the polarizability along the principal axis with geometrical factor L can be expressed as

$$\alpha \approx \frac{V}{\left(L + \frac{\varepsilon_m}{\varepsilon - \varepsilon_m}\right) + A\varepsilon_m x^2 + B\varepsilon_m^2 x^4 - i\frac{4\pi^2\varepsilon_m^{3/2}}{3}\frac{V}{\lambda_0^3}}. \tag{5.24}$$

Using the empirical data of spectra akin to Fig. 5.8, the following dependencies of A and B on L have been obtained:

$$A(L) = -0.4865L - 1.046L^2 + 0.8481L^3 \tag{5.25a}$$

$$B(L) = 0.01909L + 0.1999L^2 + 0.6077L^3 \tag{5.25b}$$

The data points in the spectra of Fig. 5.8 correspond to the extinction calculated using (5.24). We note that, perhaps surprisingly, these expressions seem to be equally valid both for gold and silver particles.

5.5 Coupling Between Localized Plasmons

We have seen that the localized plasmon resonance of a *single* metallic nanoparticle can be shifted in frequency from the Fröhlich frequency defined by (5.8) via alterations in particle shape and size. In *particle ensembles*, additional shifts are expected to occur due to electromagnetic interactions between the localized modes. For small particles, these interactions are essentially of a dipolar nature, and the particle ensemble can in a first approximation be treated as an ensemble of interacting dipoles.

We will now describe the effects of such interactions in ordered metal nanoparticle arrays. Electromagnetic coupling in disordered arrays, where interesting localization effects can occur for closely spaced particles, will be touched upon in chapter 9 when discussing enhancement processes due to field localization in particle junctions. Here, we assume that the particles of size a are arranged within ordered one- or two-dimensional arrays with interparticle spacing d. We further assume that $a \ll d$, so that the dipolar approximation is justified, and the particles can be treated as point dipoles.

Two regimes have to be distinguished, depending on the magnitude of the interparticle distance d. For closely spaced particles, $d \ll \lambda$, near-field interactions with a distance dependence of d^{-3} dominate, and the particle array can be described as an array of point dipoles interacting via their near-field (see (5.11)). In this case, strong field localization in nano-sized gaps between adjacent particles has been observed for regular one-dimensional particle chains [Krenn et al., 1999]. The field localization is due to a suppression of scattering into the far-field via excitation of plasmon modes in particles along the chain axis, mediated by near-field coupling. Fig. 5.9 illustrates this fact by showing the experimentally observed (a, c) and simulated (b, d) distribution of the electric field above single gold nanoparticles and a particle chain. In this study by Krenn and co-workers, the structures were excited using prism coupling from the substrate side and the optical near-field was probed by near-field microscopy in collection mode. From the images, it can clearly be seen that scattering is drastically suppressed for closely spaced particles, and that the fields are instead highly localized at interstitial sites. Interparticle junctions such as these therefore serve as hot-spots for field enhancement, which will be further discussed in chapter 9 in a context of surface-enhanced Raman scattering (SERS).

One can intuitively see that interparticle coupling will lead to shifts in the spectral position of the plasmon resonance compared to the case of an isolated particle. Using the simple approximation of an array of interacting point dipoles, the direction of the resonance shifts for in-phase illumination can be determined by considering the Coulomb forces associated with the polarization

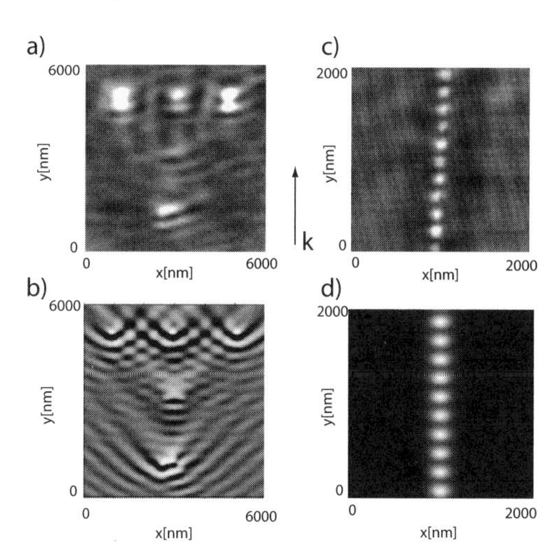

Figure 5.9. Experimentally observed (a, c) and simulated (b, d) intensity distribution of the optical near-field above an ensemble of well-separated gold particles (a, b) and a chain of closely spaced gold nanoparticles (c, d). While for separated particles interference effects of the scattered fields are visible, in the particle chain the fields are closely confined in gaps between adjacent particles. Plasmon resonances were excited using prism coupling with the direction of the in-plane moment component as outlined in the pictures. Reprinted with permission from [Krenn et al., 2001]. Copyright 2001 by Blackwell Publishing.

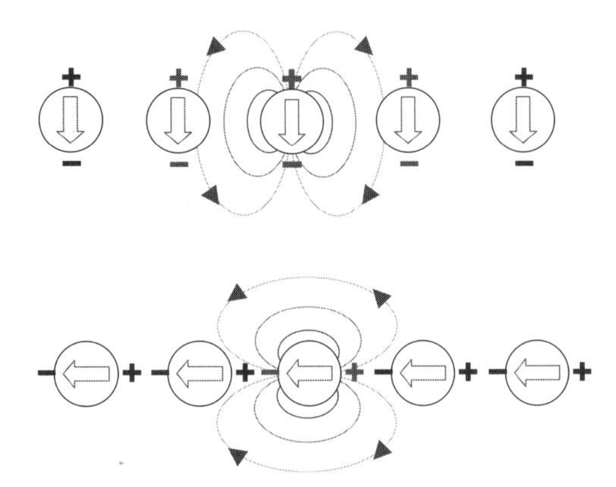

Figure 5.10. Schematic of near-field coupling between metallic nanoparticles for the two different polarizations.

of the particles. As sketched in Fig. 5.10, the restoring force acting on the oscillating electrons of each particle in the chain is either increased or decreased by the charge distribution of neighboring particles. Depending on the polarization direction of the exciting light, this leads to a blue-shift of the plasmon resonance for the excitation of transverse modes, and a red-shift for longitudinal modes.

Using one-dimensional arrays of 50 nm gold particles with varying interparticle distance (Fig. 5.11a), these shifts in resonance energy were experimentally demonstrated using far-field extinction spectroscopy [Maier et al., 2002a]. The dependence of the spectral position of the plasmon resonance on interparticle distance is shown in Fig. 5.11b both for longitudinal and transverse polarization. Due to the strong scaling of the interaction strength with d^{-3} (see (5.11)), particle separations in excess of 150 nm are sufficient to recover the behavior of essentially isolated particles.

The spatial extent of near-field interactions can further be quantified by analyzing the dependence of the resonance shifts on the length of the particle arrays [Maier et al., 2002b]. Fig. 5.12 shows results from finite-difference time-domain (FDTD) calculations and comparisons to experimental shifts obtained for chains of gold nanoparticles with fixed interparticle distance and varying chain lengths. In the FDTD simulations, the time-dependence of the electric field was monitored at the center of a particle of a chain consisting of seven 50 nm gold spheres separated by 75 nm in air (left panel). The upper inset shows the distribution of the initial electric field around the structure upon in-phase excitation with longitudinal polarization, and the lower inset the Fourier transform of the time-domain data, peaking at the longitudinal resonance frequency E_L. A comparison with chains fabricated on a silica substrate using electron beam lithography is shown in the right panel (a). As apparent, the collective plasmon resonance energies for both longitudinal (E_L) and transverse

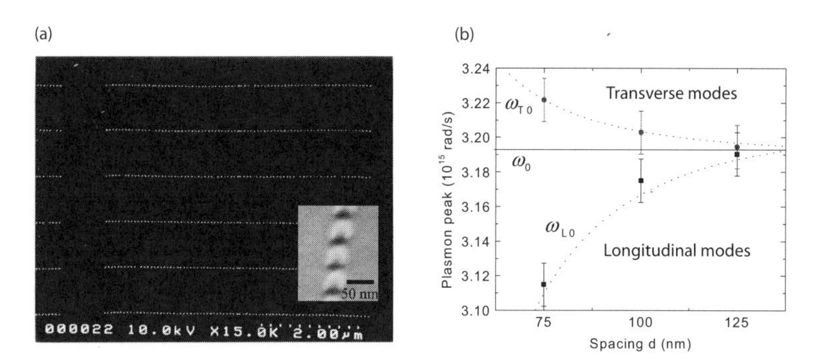

Figure 5.11. SEM image of arrays of closely spaced gold nanoparticles (a) and dependence of the spectral position of the dipole plasmon resonance on interparticle spacing (b). The dotted lines show a fit to the d^{-3} dependence of the coupling expected from a point-dipole model. Reprinted with permission from [Maier et al., 2002a]. Copyright 2002 by the American Physical Society.

(E_T) excitations for gold nanoparticle arrays of different lengths asymptote already for a chain length of about 5 particles due to the near-field nature of the coupling. The coupling strength between adjacent particles can be increased by changing the geometry to spheroidal particles (Fig. 5.12b). We point out that due to near-field interactions, a linear array of closely-spaced metal nanopar-

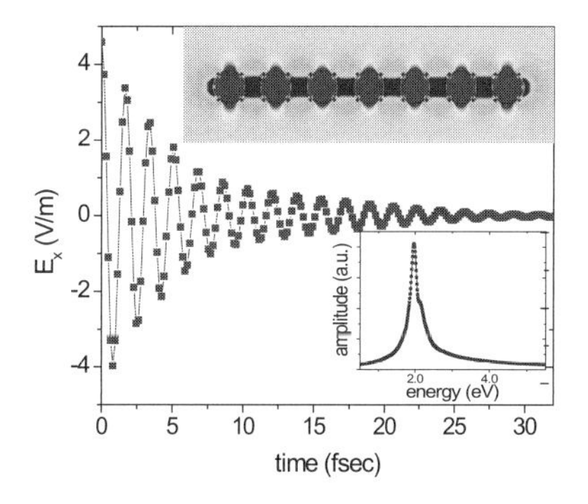

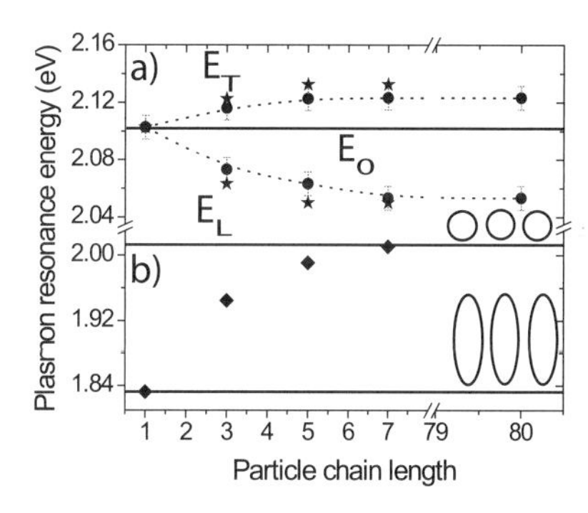

Figure 5.12. Left: Time-dependence of the electric field monitored at the center of a particle of a chain consisting of seven 50 nm gold spheres separated by 75 nm in air obtained using FDTD. For details see text. Right: a) Collective plasmon resonance energies for both longitudinal (E_L) and transverse (E_T) excitations for gold nanoparticle arrays (of the same geometry as in the left panel) obtained via far-field spectroscopy on fabricated arrays (circles) and FDTD simulations (stars). b) Simulation results for the collective plasmon resonance energies for transverse excitation of gold spheroids with aspect ratios 3:1 (diamonds). Reproduced with permission from [Maier et al., 2002b]. Copyright 2002, American Institute of Physics.

ticles can therefore be viewed as a chain of interacting dipoles, which supports traveling polarization waves. This suggests applications of metal particle chains as waveguides with high field-confinement, which will be discussed in chapter 7, together with corrections to the simple point-dipole model described here.

After these initial investigations, a number of different studies using both near- and far-field detection techniques have confirmed the distance-dependence of near-field interactions in particle arrays [Wurtz et al., 2003] as well as particle pairs [Su et al., 2003, Sundaramurthy et al., 2005]. For a detailed analysis of near-field interactions in particle ensembles of various lengths and shapes using Mie theory, we refer to the treatment by Quinten and Kreibig [Quinten and Kreibig, 1993]. Also, near-field coupling can influence plasmon resonances sustained by a single particle of complex shape, for example crescent moon structures with two sharp edges in small proximity of each other [Kim et al., 2005].

For larger particle separations, far-field dipolar coupling with a distance dependence of d^{-1} (see (5.12)) dominates. This coupling via diffraction has been analyzed for both two-dimensional arrays akin to gratings [Lamprecht et al., 2000, Haynes et al., 2003], and one-dimensional chains with interparticle distances larger than those for which near-field coupling is observed [Hicks et al., 2005]. For the example of two-dimensional gratings of gold nanoparticles with various lattice constants, Fig. 5.13 shows that far-field coupling has pronounced influences on the plasmon lineshape, both in terms of resonance frequency as well as spectral width. The latter is due to a drastic dependence of the decay time T_2 on the grating constant via its influence on the amount of radiative damping as successive grating orders change from evanescent to

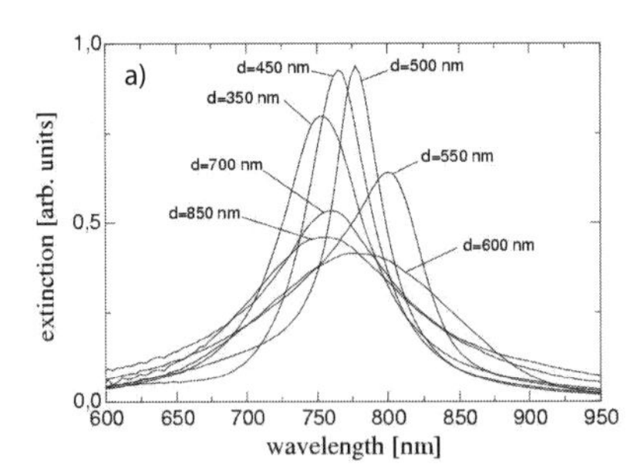

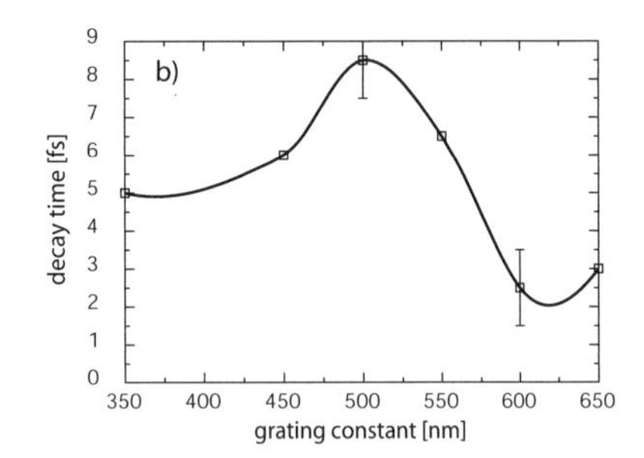

Figure 5.13. (a) Extinction spectra for square two-dimensional gratings of gold nanoparticles (height 14 nm, diameter 150 nm) with grating constant d situated on a glass substrate. (b) Plasmon decay time versus grating constant observed using a time-resolved measurement. The solid curve is a guide to the eye. Reproduced with permission from [Lamprecht et al., 2000]. Copyright 2000 by the American Physical Society.

radiative in character. In this study, the decay time of the plasmon oscillations has been determined directly using time-resolved measurements.

We note that interactions between metal nanoparticles can be further enhanced by providing additional coupling pathways, for example in the form of propagating SPPs for particle arrays fabricated on a conductive substrate [Félidj et al., 2002].

5.6 Void Plasmons and Metallic Nanoshells

We take now a step back and continue our previous discussion of plasmon resonances in single particles by taking a closer look at localized modes in metallic structures containing dielectric inclusions of characteristic dimension $a \ll \lambda$. The simplest such structure is a spherical inclusion of dielectric constant ε_m in a homogeneous metallic body described by the dielectric function $\varepsilon(\omega)$, as pictured in Fig. 5.14. Such a *nanovoid* can sustain an electromagnetic dipole resonance akin to that of a metallic nanoparticle. In fact, the result for the dipole moment of the void can be obtained from that for a sphere by simply carrying out the substitutions $\varepsilon(\omega) \rightarrow \varepsilon_m$ and $\varepsilon_m \rightarrow \varepsilon(\omega)$ in (5.7). The polarizability of the nanovoid is thus

$$\alpha = 4\pi a^3 \frac{\varepsilon_m - \varepsilon}{\varepsilon_m + 2\varepsilon}. \tag{5.26}$$

Note that contrary to metal nanoparticles, the induced dipole moment is in this case oriented antiparallel to the applied outside field. The Fröhlich condition now takes the form

$$\mathbf{Re}[\varepsilon(\omega)] = -\frac{1}{2}\varepsilon_m. \tag{5.27}$$

An important example of a three-dimensional void resonance is that of a core/shell particle consisting of a dielectric core (usually silica) and a thin metallic shell (for example gold). The polarizability of this core/shell system can be described using quasistatic Mie theory by (5.18). In an illuminating analysis, Prodan and co-workers demonstrated that the two fundamental dipolar modes of a core/shell nanoparticle can be thought to arise via the hy-

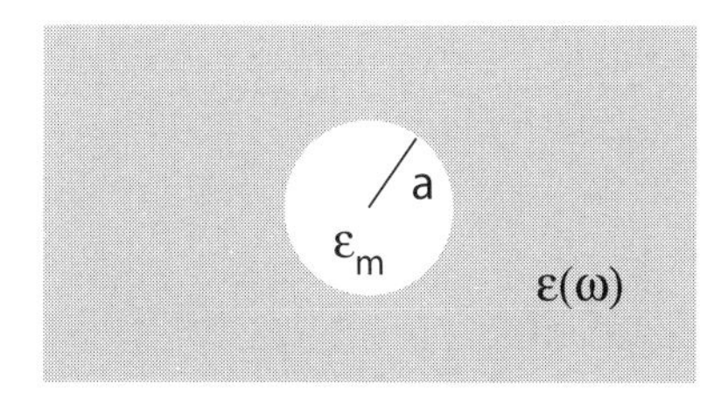

Figure 5.14. Spherical dielectric inclusion in a homogeneous metal.

bridization of the dipolar modes of a metallic sphere and a dielectric void in a metallic substrate (Fig. 5.15) [Prodan et al., 2003b]. In this picture, the two distinct nanoshell resonances are due to bonding and anti-bonding combinations of the fundamental sphere and void modes. The validity of this model has been confirmed using quantum-mechanical calculations [Prodan and Nordlander, 2003, Prodan et al., 2003a] as well as finite-difference time-domain simulations [Oubre and Nordlander, 2004].

For a quantitative description of plasmon hybridization applied to the geometry presented in Fig. 5.15, the particle plasmon can be described as an incompressible deformation of the conduction electron gas of the metallic nanostructure [Prodan et al., 2003b]. Such deformations can be expressed using spherical harmonics of order l, and as the outcome of this study, the resonance frequencies $\omega_{l,\pm}$ of the two hybridized modes for each order $l > 0$ can be written as

$$\omega_{l,\pm}^2 = \frac{\omega_\mathrm{p}^2}{2}\left[1 \pm \frac{1}{2l+1}\sqrt{1+4l\,(l+1)\left(\frac{a}{b}^{2l+1}\right)}\right], \tag{5.28}$$

where a and b are the inner and outer radius of the shell, respectively. The hybridization model has also successfully been applied to the calculation of the resonance frequencies of nanoparticle dimers [Nordlander et al., 2004].

The extra degrees of freedom over the control of the plasmon dipole resonance in the nanoshell geometry enable both shifts of the resonance frequencies into the near-infrared region of the spectrum, and additionally reduced plasmon linewidths [Teperik and Popov, 2004, Westcott et al., 2002]. The

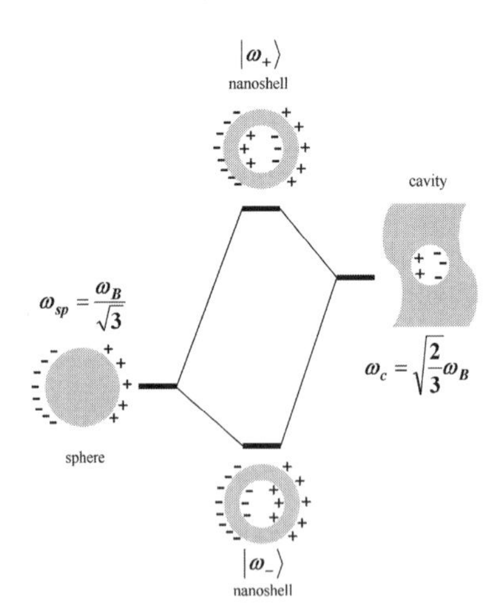

Figure 5.15. Schematic of plasmon hybridization in metallic nanoshells. Note that $\omega_B \equiv \omega_\mathrm{p}$. Reprinted with permission from [Prodan et al., 2003b]. Copyright 2003, AAAS.

latter fact suggests that nanoshells are superior to solid metal nanoparticles for refractive index sensing applications [Raschke et al., 2004, Tam et al., 2004]. Strong localized plasmon resonances in the near-infrared region of the spectrum are of interest for biomedical applications, e.g. for the treatment of nanoparticle-filled tumors, which can be destroyed via absorption-induced heating [Hirsch et al., 2003].

While the above voids have been three-dimensional in nature, essentially two-dimensional holes in thin metallic films can also support localized plasmon modes. Such structures can for example be fabricated using focused ion beam milling, and be investigated using near-field optical spectroscopy [Prikulis et al., 2004, Yin et al., 2004]. This geometry is also promising from a sensing viewpoint [Rindzevicius et al., 2005]. We will take a closer look on the fascinating properties of these systems in chapter 8.

5.7 Localized Plasmons and Gain Media

We want to finish this section by taking a brief look at an emerging application in plasmonics, namely the interaction of localized resonances with gain media. The motivation for this application is twofold: the field enhancement sustained by the metallic nanostructures upon resonant excitation can lead to a reduction in the threshold for achieving inversion in the optically active surrounding medium, and the presence of gain can counteract the inherent absorption losses in the metal. While this strengthening of plasmon resonances in gain media has up to this point not been experimentally confirmed, amplification of fluorescence due to field enhancement in mixtures of laser dyes with metal nanoparticles has recently been observed [Dice et al., 2005].

In its simplest form, the problem of a gain-induced increase in the strength of the plasmon resonance can be treated by analyzing the case of a sub-wavelength metal nanosphere embedded in a homogeneous medium exhibiting optical gain. The quasi-static approach presented at the beginning of this chapter can be followed, and the presence of gain incorporated by replacing the real dielectric constant ε_m of the insulator surrounding the sphere with a complex dielectric function $\varepsilon_2(\omega)$.

Using this straightforward analytical model, Lawandy has shown that the presence of gain, expressed by $\mathrm{Im}[\varepsilon_2] < 0$, can lead to a significant strengthening of the plasmon resonance [Lawandy, 2004]. This is due to the fact that in addition to the cancellation of the real part of the denominator of the polarizability α (5.7), the positive imaginary part of ε_2 can in principle lead to a complete cancellation of the terms in the denominator and thus to an infinite magnitude of the resonant polarizability. Taking as a starting point the expressions for the electric fields (5.9), the depolarization field $\mathbf{E}_{\mathrm{pol}} = \mathbf{E}_{\mathrm{in}} - \mathbf{E}_0$ inside the particle is given by

$$\mathbf{E}_{\text{pol}} = \frac{\varepsilon_2 - \varepsilon}{\varepsilon + 2\varepsilon_2}\mathbf{E}_0. \tag{5.29}$$

For a Drude metal with ε given by (1.20) in the small-damping limit with electron scattering rate $\gamma \ll \omega$, the incomplete vanishing of the denominator in (5.29) upon resonance can be overcome by optical gain. Ignoring gain saturation, it can be shown that the critical gain value α_c at the plasmon resonance ω_0 for the singularity to occur can be approximated as

$$\alpha_c = \frac{\gamma\,(2\text{Re}\,[\varepsilon(\omega_0)] + 1)}{2c\sqrt{\text{Re}\,[\varepsilon(\omega_0)]}}. \tag{5.30}$$

For silver and gold particles, this results in $\alpha_c \approx 10^3\ \text{cm}^{-1}$. Of course, in real examples the divergence in field amplification will be suppressed due to gain saturation, and we refer the reader to [Lawandy, 2004] for more details. Further comments on the interaction of gain media with plasmons in a context of waveguiding will be presented in chapter 7.

Chapter 6

ELECTROMAGNETIC SURFACE MODES AT LOW FREQUENCIES

We have seen in previous chapters that surface plasmon polaritons can confine electromagnetic fields to the interface between a dielectric and a conductor over length scales significantly smaller than the wavelength. This high field localization occurs as long as the fields oscillate at frequencies close to the intrinsic plasma frequency of the conductor. The most promising applications of plasmonics based on *metals*, such as highly localized waveguiding and optical sensing with unprecedented sensitivity (which will be discussed in part II of this book), have therefore been limited to the visible or near-infrared part of the spectrum. At lower frequencies, a brief look at the SPP dispersion relation reveals that the confinement to the interface breaks down as the propagation constant rapidly decreases towards the wave vector in the dielectric.

Therefore, for typical metals such as gold or silver, SPPs evolve into grazing incidence light fields as the frequency is lowered, extending over a great number of wavelengths into the dielectric space above the interface. The underlying physics of this evolution from a highly confined surface excitation to an essentially homogeneous light field in the dielectric, propagating along the interface with the same phase velocity as unbound radiation, is the decrease in field penetration into the conductor at lower frequencies, due to the large (negative) real and (positive) imaginary parts of the permittivity. Since an appreciable field amplitude inside the metal is essential for providing the non-zero component of the electric field parallel to the surface necessary for the establishment of an oscillating spatial charge distribution, SPPs vanish in the limit of a *perfect electrical conductor*. Highly doped *semiconductors* however can exhibit plasma frequencies at mid- and far-infrared frequencies, and thus allow SPP propagation akin to metals at visible frequencies, albeit with high losses.

Taking the technologically important THz spectral regime ($0.5\ \mathrm{THz} \leq f \leq 5\ \mathrm{THz}$) as an example, this chapter first briefly examines the propagation of

SPPs at flat metal or semiconducting interfaces. We then show that even perfect conductors can support electromagnetic surface waves closely resembling SPPs provided that the surface is textured. These *designer* or *spoof plasmons* show a rich physics and could have a number of important applications, specifically for highly sensitive biological sensing and near-field imaging using THz waves. While not directly related to plasmonics, the chapter closes with a short look at surface *phonon* polaritons, coupled excitations of the electromagnetic field and phonon modes of polar materials such as SiC occurring at mid-infrared frequencies.

6.1 Surface Plasmon Polaritons at THz Frequencies

As discussed in detail in chapter 2, the localization and concomitant field enhancement offered by SPPs at the interface between a conductor and a dielectric with refractive index n is due to a large SPP propagation constant $\beta > k_0 n$, leading to evanescent decay of the fields perpendicular to the interface. The amount of confinement increases with β according to (2.13). Conversely, localization significantly decreases for frequencies $\omega \ll \omega_p$, where $\beta \to k_0 n$.

Due to their large free electron density $n_e \approx 10^{23}$ cm^{-3}, metals support *well-confined* SPPs only at visible and near-infrared frequencies. As shown in Fig. 6.1 for the example of a silver/air interface, $\beta \approx k_0$ at far-infrared frequencies in the THz regime, in fact to an accuracy of about 1 part in 10^5. This is due to the large complex permittivity $|\varepsilon| \approx 10^5$, leading to negligible field penetration into the conductor and thus highly delocalized fields. For metals, SPPs at these frequencies therefore nearly resemble a homogeneous light field in air incident under a grazing angle to the interface, and are also known as *Sommerfeld-Zenneck waves* [Goubau, 1950, Wait, 1998]. We note

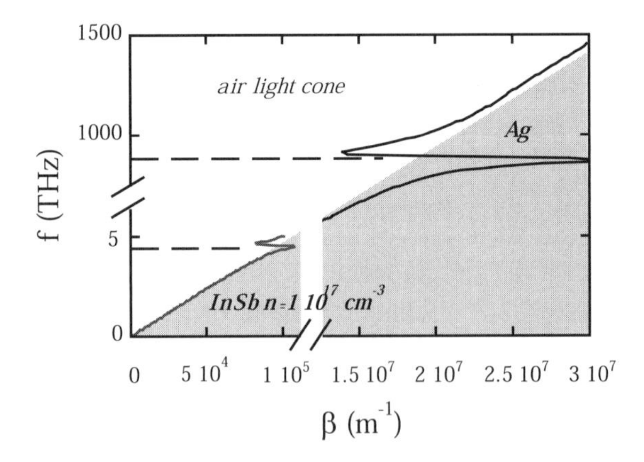

Figure 6.1. SPP dispersion relation for a flat silver/air and InSb/air interface (courtesy of Steve Andrews, University of Bath).

that all expressions derived in the discussion of SPPs at visible frequencies in chapter 2 are also valid in the low-frequency regime if the appropriate dielectric data for metals is used, for example those obtained in [Ordal et al., 1983].

Fig. 6.1 also shows the SPP dispersion relation for the interface between air and a highly doped semiconductor, in this case InSb with $n_e \approx 10^{17}$ cm^{-3}. As can be seen, due to the lower free electron density, such semiconductors can exhibit a SPP propagation constant $\beta > k_0 n$ and thus field-localization at THz frequencies resembling that for metals at visible frequencies, however with accompanying large absorption. Plasmon propagation of broadband THz pulses at the interface of a highly doped silicon grating has indeed been observed [Gómez-Rivas et al., 2004]. One intriguing aspect of using semiconductors for low-frequency SPP propagation apart from the enhanced confinement is the possibility to tune the carrier density and thus ω_p by either thermal excitation, photocarrier generation or direct carrier injection. Thus, active devices for switching applications seem possible. As a first step in this direction, Gómez-Rivas and co-workers have demonstrated the modification of Bragg scattering of THz SPPs on a InSb grating using thermal tuning [Gómez-Rivas et al., 2006]. We will in the following mostly focus on metals however, because of the interesting possibility to engineer the dispersion of surfaces waves at will using a geometry-based approach.

The excitation and detection of broadband THz pulses, also known as THz time-domain spectroscopy, usually employs a coherent generation and detection scheme [van Exter and Grischkowsky, 1990]. This allows a direct investigation of both amplitude and phase of the propagating SPPs. A typical setup is

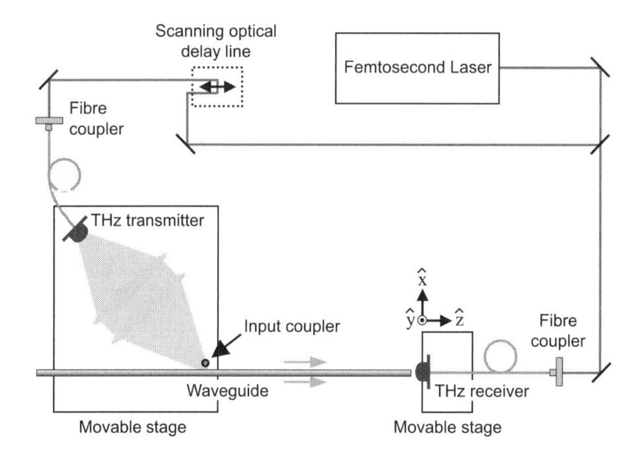

Figure 6.2. Typical setup for the generation and detection of broadband THz pulses. Input coupling to SPPs is achieved using scattering at a small gap between the guiding structure and a sharp edge. Reprinted with permission from Macmillan Publishers Ltd: Nature [Wang and Mittleman, 2005], copyright 2004.

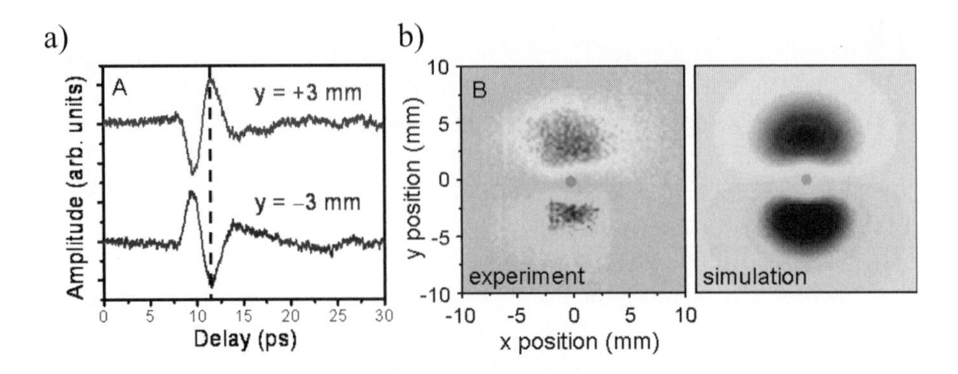

Figure 6.3. THz SPP propagation on a bare stainless-steel wire. a) Time-domain electric field waveform 3 mm above and below the wire. b) Experimental and simulated spatial mode profile. The radial nature of the mode is evident. Reprinted with permission from Macmillan Publishers Ltd: Nature [Wang and Mittleman, 2005], copyright 2004.

shown in Fig. 6.2. Short light pulses generated by a femtosecond laser are split into two lightpaths using a semitransparent mirror. The pulses propagating in the generation pathway create photocarriers in a THz transmitter consisting of two biased electrodes on a semiconductor substrate, leading to a current surge between the electrodes and the radiation of THz waves. Conversely, the pulses in the detection pathway are used for photocarrier generation in the unbiased receiver, and sampling of the THz waveform is enabled by introducing a variable time delay between the two pathways. Conversion of a fraction of the power carried by the generated free-space THz pulse into SPPs is conveniently accomplished using *edge* or *aperture coupling*: the pulse is focused on a small gap of size of the order of or smaller than the wavelength ($\lambda \approx 300\ \mu$m at 1 THz) between a razor blade and the structure supporting SPPs. Scattering at this edge provides the additional wave vector components necessary for phase-matching, albeit with generally low efficiency.

The propagation of THz SPPs on flat metal films has been investigated using these broadband techniques, confirming the highly delocalized nature of the modes. For example, their penetration into the air space above a gold film up to distances of multiple centimeters has been demonstrated for frequencies around 1 THz [Saxler et al., 2004]. We note that the slow decay of the wave into the dielectric medium is not the only consequence of $\beta = k_0 n$. Also, the phase velocity of the surface waves is equal to that of the waves propagating in free space used to excite the pulse. Therefore, power can be transfered back and forth between the two waves if they are allowed to co-propagate along the interface, which makes the detailed investigation of THz SPPs challenging. This is highlighted by the fact that discrepancies on the order of 1-2 magnitudes between the spatial extent and attenuation length predicted from theory and experimental investigations have been reported for THz SPPs propagating on a thin aluminum sheet [Jeon and Grischkowsky, 2006]. An explanation of this fact could lie in the difficulties in exciting a *pure* Sommerfeld-Zenneck wave, due to its highly unconfined nature.

In addition to flat films, also cylindrical structures such as metallic wires can efficiently guide delocalized THz SPPs. Using a typical time-domain spectroscopy setup (Fig. 6.2), Wang and Mittleman investigated the propagation of SPPs on a thin stainless-steel wire [Wang and Mittleman, 2005] and demonstrated the potential usefulness of this simple geometry for practical applications in THz waveguiding technology. In this study, an attenuation constant of only $\alpha = 0.03\ \mathrm{cm}^{-1}$ has been determined, and the radial nature of the mode confirmed. This is illustrated in Fig. 6.3, which compares the mode profile determined via sampling of the time-domain electric field waveforms around the wire with the mode profile expected from Sommerfeld theory [Goubau, 1950]. The agreement between the theoretically and experimentally obtained intensity distributions has been corroborated in further studies [Wachter et al., 2005]. Apart from low damping, $\beta = k_0$ further leads to an extremely low group velocity dispersion, allowing essentially undistorted pulse propagation. However, a detrimental consequence of the highly delocalized nature of the propagating modes are significant radiation losses at bends [Jeon et al., 2005] or irregularities, limiting practical applications.

Recent studies have also revealed that *localized* plasmons can be excited at THz frequencies. For example, micron-sized silicon particles support dipolar plasmon resonances akin to the Fröhlich modes presented in chapter 5, with a frequency depending on the concentration of free carriers n_e due to the scaling $\omega_\mathrm{p} \propto \sqrt{n_\mathrm{e}}$ [Nienhuys and Sundström, 2005]. Localized modes have also been observed in ensembles of randomly distributed metallic particles in the context of enhanced transmission of THz radiation [Chau et al., 2005]. Since the physics of the localization process is essentially equal to that discussed for nanoparticles at optical frequencies, we will not embark on a detailed discussion.

6.2 Designer Surface Plasmon Polaritons on Corrugated Surfaces

We have seen that due to the large permittivity of metals at THz frequencies, SPPs in this regime are highly delocalized. Physically, this is due to the negligible field penetration into the metal - only a vanishingly small fraction of the total electric field energy of the SPP mode resides inside the conductor. In the limit of a perfect conductor, the internal fields are identically zero. Perfect metals thus do not support electromagnetic surface modes, forbidding the existence of SPPs.

However, Pendry and co-workers have shown that bound electromagnetic surface waves mimicking SPPs can be sustained even by a perfect conduc-

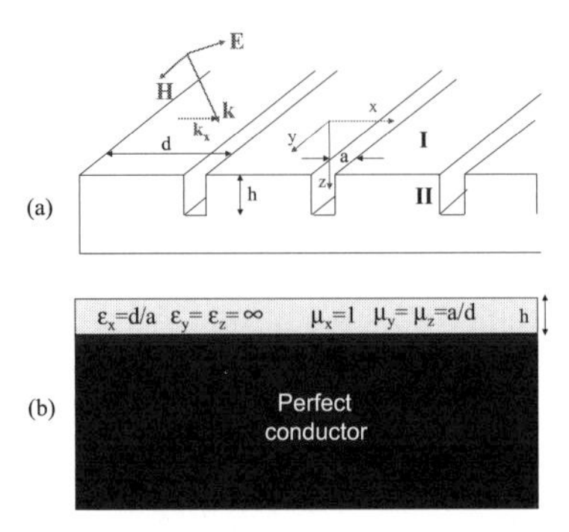

Figure 6.4. (a) One-dimensional array of grooves of width a and depth h with lattice constant d. (b) The effective medium approximation of the structure. Reprinted with permission from [García-Vidal et al., 2005a]. Copyright 2005, Institute of Physics.

tor, provided that its surface is periodically corrugated [Pendry et al., 2004]. For real metals with finite conductivity, these *designer* or *spoof* SPPs should dominate over the delocalized Sommerfeld-Zenneck waves. If the size and spacing of the corrugations is much smaller than the wavelength λ_0, the photonic response of the surface can be described by an *effective medium* dielectric function $\varepsilon(\omega)$ of the plasma form, with ω_p determined by the geometry. Thus, the dispersion relation of the surface mode can be engineered via the geometry of the surface, allowing tailoring to particular frequencies. In the effective medium model, the establishment of surface waves can be physically understood by realizing that the surface modulations allow for an average *finite* field penetration into the effective surface layer, akin to the field penetration into real metals at visible frequencies leading to the formation of confined SPPs. A material with sub-wavelength structure exhibiting such an effective photonic response is also known as a *metamaterial*.

While it can be shown that any periodic modulation of the flat surface of a perfect conductor will lead to the formation of bound surface states, we present here two prominent geometries, closely following the reasoning and notation by García-Vidal and co-workers [García-Vidal et al., 2005a] - a one-dimensional array of grooves and a two-dimensional hole array. The approach here should be generally applicable for the investigation of surface modes. We note that the frequencies of the supported modes scale with the geometrical size of the corrugations in the perfect conductor approximation.

Fig. 6.4a shows the geometry of a one-dimensional array of grooves of width a and depth h separated by a lattice constant d on the surface of a perfect conductor. The dispersion relation $\omega(k_x)$ of the surface mode with propagation constant $k_x = \beta$ sustained by the modulated interface can be calculated by examining the reflectance of a TM-polarized incident wave. The reasoning be-

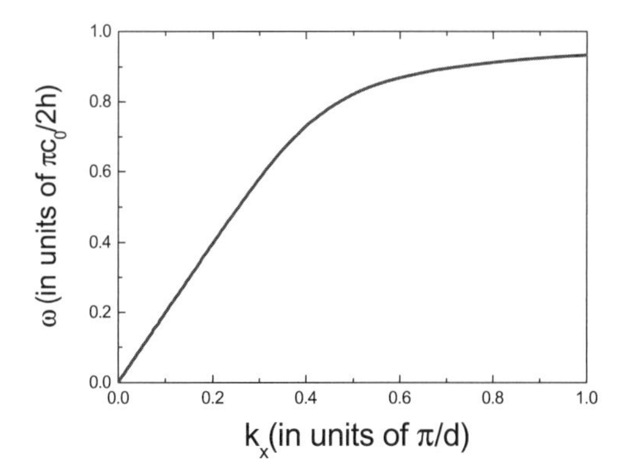

Figure 6.5. Dispersion relation (6.4) of the SPP-like mode of a groove array for the case $a/d = 0.2$ and $h = d$. Reprinted with permission from [García-Vidal et al., 2005a]. Copyright 2005, Institute of Physics.

hind this technique is that the surface mode resonance corresponds to a divergence in the reflectivity - the mode can exist for a vanishingly small excitation. For the calculation of the reflectivity, the total field above the surface in the vacuum region is written as a sum of the incident field and the reflected fields of diffraction order n, and the fields inside the grooves are expanded into the fundamental forward and backward propagating TE-modes (in the z-direction perpendicular to the surface). The restriction to the fundamental TE-mode is valid for $\lambda_0 \gg a$, i.e. a groove width much smaller than the free-space wavelength. By matching the appropriate boundary conditions for the electric and magnetic fields, the reflection coefficients for diffraction order n calculate to

$$\rho_n = -\frac{2i \tan(k_0 h)\, S_0 S_n k_0/k_z}{1 - i \tan(k_0 h) \sum_{n=-\infty}^{\infty} S_n^2 k_0/k_z^{(n)}}, \tag{6.1}$$

where $k_0 = \omega/c$ and $k_z^{(n)} = \sqrt{k_0^2 - \left(k_x^{(n)}\right)^2}$ with $k_x^{(n)} = k_x + 2\pi n/d$ for diffraction order n. S_n is the overlap integral between the nth-order plane wave and the fundamental TE mode and evaluates to

$$S_n = \sqrt{\frac{a}{d}} \frac{\sin\left(k_x^{(n)} a/2\right)}{k_x^{(n)} a/2}. \tag{6.2}$$

The dispersion relation of surface modes is now determined by the poles of the reflection coefficients (6.1). Assuming that $\lambda_0 \gg d$ so that only the specular reflection order with coefficient ρ_0 needs to be taken into account, and additionally that $k_x > k_z$ (since we are interested in a mode confined to the surface), the dispersion relation of the bound state can be expressed as

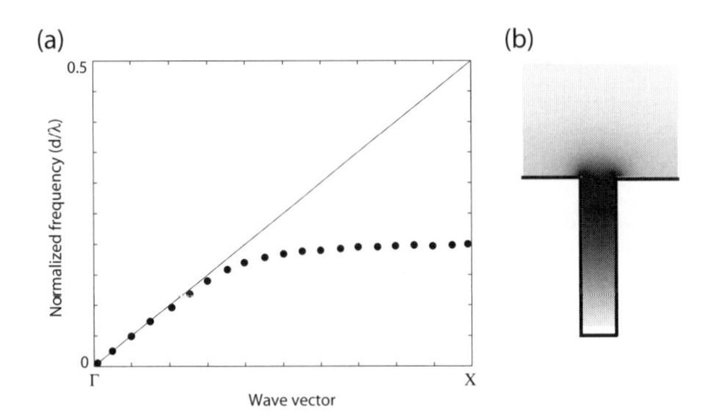

Figure 6.6. Dispersion relation (a) and electric field distribution in the unit cell at the band edge (b) for a SPP-like surface mode for $h = d = 50\mu$m and $a = 10\mu$ calculated using finite-difference time-domain simulations.

$$\frac{\sqrt{k_x^2 - k_0^2}}{k_0} = S_0^2 \tan k_0 h. \tag{6.3}$$

This relation is valid for $\lambda_0 \gg a, d$ (effective medium approximation).

The similarity of the excitations described by (6.3) to SPPs can be elucidated by relating the dispersion relation to that of electromagnetic waves at the surface of a homogeneous *anisotropic* dielectric of height h on top of the perfectly conducting substrate (Fig. 6.4b). If we define its permittivity such that $\varepsilon_x = d/a$, $\varepsilon_y = \varepsilon_z = \infty$, a straight-forward analysis of light propagation inside the grooves shows that the corresponding magnetic permeability is $\mu_x = 1$ and $\mu_y = \mu_z = a/d$. Using a similar analysis of the reflection coefficient as presented above, the dispersion relation of the surface mode of this anisotropic structure is

$$\frac{\sqrt{k_x^2 - k_0^2}}{k_0} = \frac{a}{d} \tan k_0 h, \tag{6.4}$$

which corresponds to (6.3) for $k_x a \ll 1$.

Fig. 6.5 shows a plot of (6.4) for $a/d = 0.2$ and $h = d$. As can be seen, the dispersion curve is similar to that of a SPP at the interface between a dielectric and a real metal. However, ω_p is determined by the surface geometry: For large k_x, the angular frequency approaches $\omega \to \pi c/2h$. In order to physically interpret the formation of this surface mode, we note that this frequency corresponds to that of the fundamental cavity waveguide mode inside the groove in the limit $a/d \to 0$. These resonances arise due to interference between modes propagating in the forward and backward z-direction. The surface mode is then established due to coupling between cavity modes localized in individual grooves.

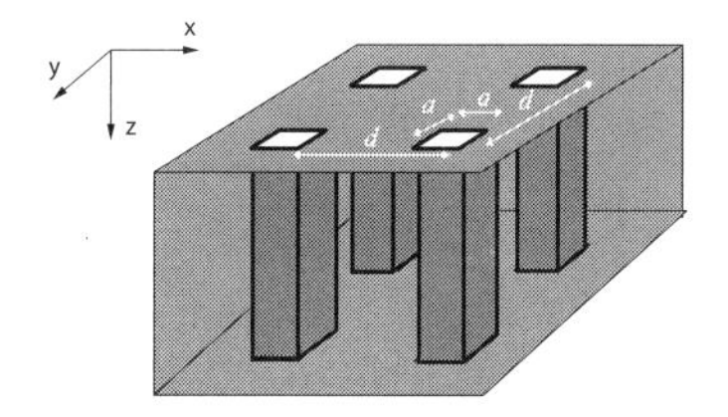

Figure 6.7. Two-dimensional square lattice of square holes of size a with lattice constant d in a semi-infinite perfect conductor. Reprinted with permission from [García-Vidal et al., 2005a]. Copyright 2005, Institute of Physics.

A more exact calculation of the dispersion relation and mode profile of designer surface plasmons supported by this geometry is provided by finite-difference time-domain calculations. As an example, Fig. 6.6 shows the dispersion (a) and mode profile (b), i.e. the distribution of $|\mathbf{E}|$, of the SPP-like surface mode for a groove array on a perfect conductor with $h = d = 50\mu$m and $a = 10\mu$m. The mode profile shows the distribution of the electric field for the surface mode at the band edge $k_x = \pi/d$. Note the high confinement of the mode to the surface. We point out that as long as both $a, d \ll \lambda_0$, the agreement between the quasi-analytical theory outlined here and finite-difference time-domain calculations is remarkable.

The second structure examined by García-Vidal and co-workers is a square lattice of square holes of side a with lattice constant d milled into a flat film (Fig. 6.7). We start by analyzing a semi-infinite structure with infinite hole depth h. The holes are filled with a non-absorbing dielectric of relative permittivity ε_h. In analogy to the discussion above, the surface modes emerge at the divergences of the reflection coefficient of a TM-polarized wave incident on the perforated surface. In the long-wavelength limit $\lambda_0 \gg d$, only the specular reflection has to be taken into account, and if we additionally impose $\lambda_0 \gg a$ so that the fundamental (decaying) eigenmode inside the holes dominates (all other modes decay much more strongly), the specular reflection coefficient ρ_0 evaluates to

$$\rho_0 = \frac{k_0^2 S_0^2 - q_z k_z}{k_0^2 S_0^2 + q_z k_z}, \tag{6.5}$$

where $q_z = \sqrt{\epsilon_h k_0^2 - \pi^2/a^2}$ is the propagation constant of the fundamental mode inside the holes and S_0 its overlap integral with the incident plane wave. Explicitly,

$$S_0 = \frac{2\sqrt{2}a \sin(k_x a/2)}{\pi d k_x a/2}. \tag{6.6}$$

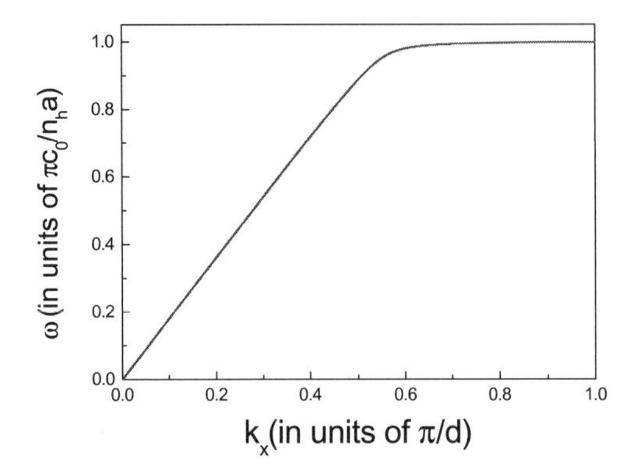

Figure 6.8. Dispersion relation (6.9) of the SPP-like bound mode at the interface between a perforated perfect conductor and vacuum for $a/d = 0.6$ and $\varepsilon_h = 9$. Reprinted with permission from [García-Vidal et al., 2005a]. Copyright 2005, Institute of Physics.

By examining the divergence of ρ_0 for $k_x > k_z$, the dispersion relation of the SPP-like bound modes evaluates to

$$\frac{\sqrt{k_x^2 - k_0^2}}{k_0} = \frac{S_0^2 k_0}{\sqrt{\pi^2/a^2 - \varepsilon_h k_0^2}}. \tag{6.7}$$

As in the discussion of the one-dimensional groove array, (6.7) can be shown to correspond to that of a homogeneous effective anisotropic semi-infinite layer in the long wavelength limit $k_x a \ll 1$. An analysis of the reflection coefficient reveals that for this system $\varepsilon_z = \mu_z = \infty$, $\mu_x = \mu_y = S_0^2$ and

$$\epsilon_x = \varepsilon_y = \frac{\varepsilon_h}{S_0^2}\left(1 - \frac{\pi^2 c_0^2}{a^2 \varepsilon_h \omega^2}\right), \tag{6.8}$$

which is of the form (1.22) with an effective plasma frequency $\omega_{\mathrm{p}} = \pi c/\sqrt{\epsilon_h} a$. It is illuminating to point out that this is the cut-off frequency of a perfect metal waveguide of square cross section with side a filled with a dielectric material of relative permittivity ε_h. Below this frequency, the electromagnetic field is exponentially decaying inside the holes, which is here the requirement for the existence of the surface state.

The dispersion relation of the surface mode supported by the interface between this effective medium and vacuum can be calculated by inserting (6.8) into expression (2.12), relating the perpendicular wave vector components k_z on both sides of the interface. We obtain

$$\frac{\sqrt{k_x^2 - k_0^2}}{k_0} = \frac{8a^2 k_0}{\pi^2 d^2 \sqrt{\pi^2/a^2 - \epsilon_h k_0^2}}, \tag{6.9}$$

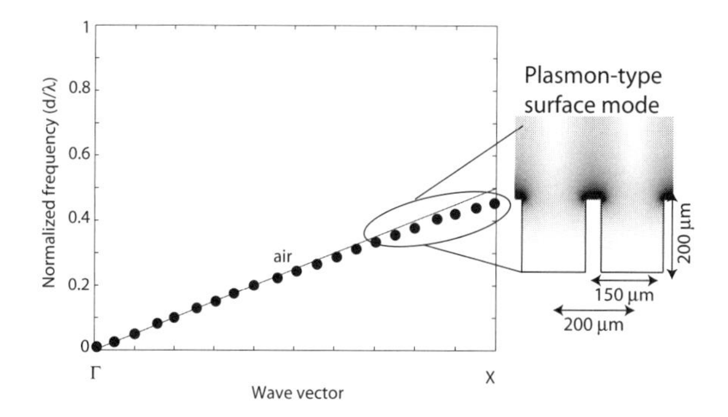

Figure 6.9. Dispersion relation of the SPP-like mode of a prefect conductor perforated with square holes of side $a = 150\mu$m and depth $h = 200\mu$m arranged on a square lattice with lattice constant $d = 200\mu$m. The distribution of the electric field at the band edge is also shown.

which is equal to (6.7) for $k_x a \ll 1$. Fig. 6.8 shows a plot of (6.9) for the case $a/d = 0.6$ and $\varepsilon_h = 9$. The size of the holes determines the amount of confinement - the smaller the holes are, the closer the dispersion will lie to the light line.

The procedure for calculating (6.9) can be extended to the case of finite holes of depth h in a straightforward manner by considering both the forward and the backward decaying modes inside the holes. The dispersion relation for this case is given by

$$\frac{\sqrt{k_x^2 - k_0^2}}{k_0} = \frac{8a^2 k_0}{\pi^2 d^2 \sqrt{\pi^2/a^2 - \varepsilon_h k_0^2}} \frac{1 - \mathrm{e}^{-2|q_z|h}}{1 + \mathrm{e}^{-2|q_z|h}} \tag{6.10}$$

with $q_z = i\sqrt{\pi^2/a^2 - \epsilon_h k_0^2}$ as above. For vanishing depth $h \to 0$, the bound mode disappears as $(k_x \to k_0)$, and for infinite depth $h \to \infty$ (6.10) evolves into (6.9). We point out that corrections to (6.9) for the long-wavelength region of the dispersion close to the light line have been suggested, due to effects of non-locality of the dielectric response [de Abajo and Sáenz, 2005]. However, as in the case of one-dimensional grooves discussed above, as long as the effective medium approximation is justified, also in this case the agreement with finite-difference time-domain simulations is very good.

In addition to this fundamental mode, for sufficient but finite hole depth h, confined surface modes with low group velocity (akin to coupled cavity modes) above the cut-off frequency ω_p for propagating modes inside the cavity holes can exist, due to the excitation of cavity resonances. These modes penetrate deeply into the holes [Qiu, 2005].

We want to stress that the theory as presented here is only valid in the limit $\lambda_0 \gg d$ and $\lambda_0 \gg a$, due to the fact that only the lowest order mode inside the holes is taken into account. For a hole size and lattice spacing not fulfilling the

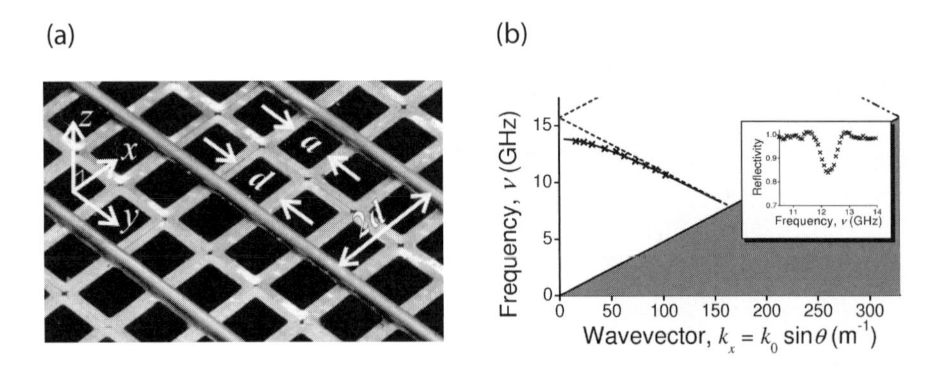

Figure 6.10. (a) Photograph of a two-dimensional array of hollow square brass tubes with side length $d = 9.525$ mm, inner dimension $a = 6.960$ mm and depth $h = 45$ mm covered with a one-dimensional array of cylindrical rods used for diffractive coupling and zone-folding. (b) Dispersion relation of the surface mode obtained via the angle-dependence of the reflectivity dips (see inset). Reprinted with permission from [Hibbins et al., 2005]. Copyright 2005, AAAS.

requirements of the effective medium approximation, finite-difference time-domain simulations are a convenient method to calculate the dispersion, taking into account the decay of higher order modes inside the holes and radiative losses. As an example, Fig. 6.9 shows the dispersion relation and mode profile of the surface modes of a perfect conductor perforated with square holes of size $a = 150\mu$m and depth $h = 200\mu$m on a square lattice with $d = 200\mu$m.

The ability to engineer the dispersion of the plasmon-like surface state of a metal interface via modulations is not restricted to perfect conductors alone. Also, for real metals the introduction of modulations will lead to a lowering of the effective plasmon frequency ω_{p} via increasing penetration of the mode into the effective surface layer. This opens up the possibility of creating structured surfaces with functional components such as waveguides or lenses by varying the refractive index $n_{\mathrm{spp}} = k_x c/\omega_{\mathrm{sp}}$ of the SPP-like mode in a controlled manner.

Hibbens and co-workers experimentally demonstrated designer plasmon surface modes supported by a two-dimensional hole array in the microwave regime using periodically arranged hollow brass tubes [Hibbins et al., 2005] (Fig. 6.10a). The existence of the surface mode was established via a study of the angle dependence of the microwave reflectivity of the structure, which allowed the determination of the dispersion relation of the surface modes via the observation of angle-dependent reflectivity dips (Fig. 6.10 inset). The introduction of a one-dimensional layer of cylindrical rods spaced by a distance $2d$ facilitated diffractive coupling and lead to zone-folding of the surface modes back into the radiative region. In Fig. 6.10b, the dispersion is therefore modified from the canonical form (6.10) via zone folding at $k_x = \pi/2d$. The surface nature of the observed mode (which is below the first-order diffracted light line associated with the rod array) is clearly confirmed.

We note that apart from planar guiding, designer SPPs also play an important role in the enhanced transmission through hole arrays for hole sizes below the cut-off of the propagating mode [Hibbins et al., 2006], which will be discussed in more detail in chapter 8.

6.3 Surface Phonon Polaritons

We have seen that at low frequencies a strong localization of the electromagnetic field with metallic structures can only be achieved for corrugated surfaces in the form of designer plasmons. While enabling sub-wavelength scale confinement even for flat surfaces, the use of conductors with lower carrier densities such as doped semiconductors suffers from the problem of high attenuation due to inherent material absorption. In this section we briefly present an interesting third option for field confinement and enhancement, which is particularly amenable to frequencies in the mid-infrared: *surface phonon polaritons*.

Surface phonon polaritons arise due to the coupling of the electromagnetic field to lattice vibrations of polar dielectrics at infrared frequencies. The physics of these excitations is conceptually similar to that of both propagating and localized surface plasmons, and the formulas derived in chapters 2 and 5 apply.

Let us give a couple of examples of both localized and propagating surface phonons. Fig. 6.11 shows a comparison of the calculated enhancement of the electric field at the Fröhlich resonance frequency for three 10 nm spheres: one consisting of SiC , and two noble metal (gold, silver) spheres [Hillenbrand et al., 2002]. It is apparent that the localized phonon resonance, situated around

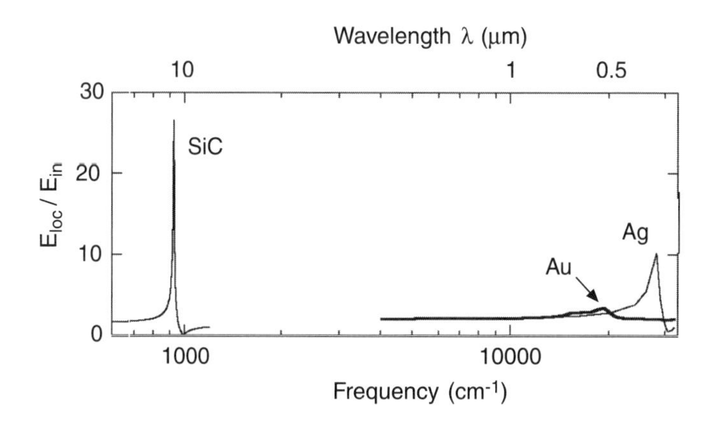

Figure 6.11. Calculated field enhancement of the polarizability of a 10 nm SiC sphere at the Fröhlich frequency defined by (5.8), compared to spheres consisting of gold or silver. Reprinted by permission from McMillan Publishers Ltd: Nature [Hillenbrand et al., 2002], copyright 2002.

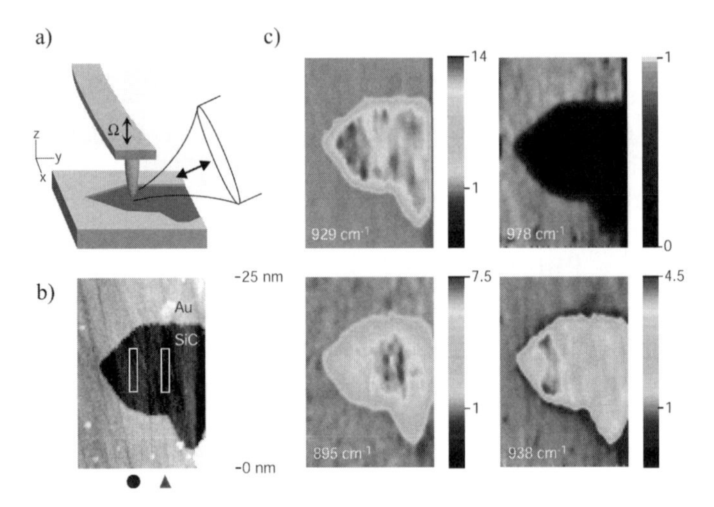

Figure 6.12. Experimental setup (a) and images (b, c) of a SiC structure surrounded by a gold film obtained with an apertureless near-field optical microscope working the mid-infrared. The topography is shown in panel (b), and near-field images in panel (c). The strong dependence of the optical contrast of the central SiC structure on wavelength is due to a resonant interaction with the probing tip at 929 cm^{-1}. Reprinted by permission from McMillan Publishers Ltd: Nature [Hillenbrand et al., 2002], copyright 2002.

a wavelength $\lambda \approx 10\ \mu\text{m}$, is significantly stronger than those of localized plasmons in the noble metals, due to the lower damping: $\text{Im}\,[\varepsilon]$ is smaller for SiC compared to gold or silver at the resonance frequency.

This suggests that photonics with phonons at mid-infrared frequencies is a promising route to sub-wavelength energy localization, in the same way as plasmonics at visible and near-infrared frequencies, with potentially smaller energy attenuation in waveguides and larger field enhancement in resonator structures. As an example of localized resonance probing, Fig. 6.12 shows the topography and near-field images of a thin SiC film surrounded by a flat gold film probed using scattering of mid-infrared radiation from a sharp platinum tip scanned over the structure [Hillenbrand et al., 2002]. As apparent, the intensity of the SiC region depends drastically on the wavelength of illumination, which is due to a resonant near-field interaction process between the structure and the tip [Renger et al., 2005].

In addition to the examination of localized phonon resonances, this technique can also be used for near-field optical imaging of surface phonon polaritons propagating on a SiC film [Huber et al., 2005]. In a study by Huber and co-workers, the propagating surface waves were excited using coupling to a free-space beam at the edge of a thin gold overlayer (Fig. 6.13a). The evanescent tail of the surface waves interacted with the probe tip, leading to scattering into the far-field and the establishment of an interference pattern (Fig. 6.13b).

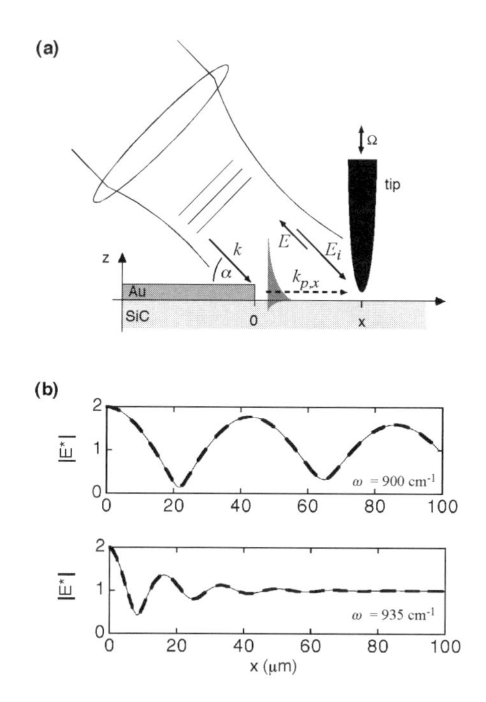

Figure 6.13. (a) Experimental setup for near-field imaging of propagating surface phonon polaritons traveling on SiC excited via edge coupling. (b) Calculated dependence of the interference pattern on excitation wavelength. Reprinted with permission from [Huber et al., 2005]. Copyright 2005, American Institute of Physics.

The use of a phase-sensitive detection technique enabled the determination of the propagation constant β and the attenuation length L via an examination of

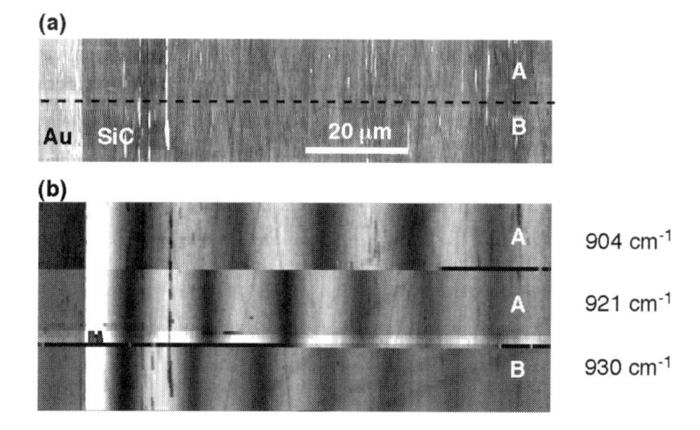

Figure 6.14. Topography (a) and near-field optical images (b) for surface phonon polariton propagation acquired using the setup of Fig. 6.13. Reprinted with permission from [Huber et al., 2005]. Copyright 2005, American Institute of Physics.

the dependence of the periodicity in intensity contrast in the obtained images on the illumination wavelength. Representative images are shown in Fig. 6.14. In this study, propagation lengths 30 μm $\leq L \leq$ 200 μm were achieved, varying with the level of confinement. It has further recently been shown that the propagation can be tailored by small changes in surface topography [Ocelic and Hillenbrand, 2004].

After these promising studies, one can expect that concepts borrowed from plasmonics at visible frequencies can and will be successfully applied in the mid-infrared using phonon excitations.

PART II

APPLICATIONS

Introduction

Armed with a sufficient background knowledge, this second part presents five prominent research areas in plasmonics. We start with an overview of different efforts to exert control over the propagation of surface plasmon polaritons. The promise behind plasmon waveguides is a new, highly integrated photonic infrastructure that might close the size gap with electronic devices. Control over the transmission of light through sub-wavelength apertures via plasmon excitations is an equally exciting area, which has spurred a tremendous amount of research ever since the initial description in 1998 of enhanced transmission of light encountered in aperture arrays. The next two chapters describe how highly localized fields around metallic nanostructures can lead to dramatic enhancements of the emission of molecules placed into these hot spots, and different methods for spectroscopy of localized modes. These chapters also include a cursory discussion of biological sensing and labeling using surface plasmons. We close with a short introduction into the field of metallic metamaterials, artificial constructs with sub-wavelength structure that exhibit novel optical phenomena such as artificial magnetism or indeed a negative refractive index.

Chapter 7

PLASMON WAVEGUIDES

Having described the basics of surface plasmon polaritons in chapter 2, we continue the discussion by providing a number of examples of control over their propagation in the context of waveguiding. Here, the trade-off between confinement and loss demands a judicious choice of geometry, depending on the length scale over which energy is to be transferred. For example, thin metallic slabs embedded in a homogeneous dielectric medium can guide SPPs over distances of many centimeters at near-infrared frequencies, but the associated fields are only weakly confined in the perpendicular direction. In the other extreme, metal nanowire or nanoparticle waveguides exhibit a transverse mode confinement below the diffraction limit in the surrounding host, but with large attenuation losses, leading to propagation lengths on the order of micrometers or below.

Routing of SPPs on planar interfaces can be achieved by locally modifying their dispersion via surface modulations, which will be described in the first two sections of this chapter. We then focus on studies of lateral confinement in metal stripe and wire waveguides, including focusing of SPPs in conical structures. The inverse structure to metal stripes, namely metal/insulator/metal heterostructures, also show high promise for waveguiding with good confinement and acceptable propagation length, especially in V-groove geometries. Towards the end of this chapter, we show that localized plasmon excitations in metal nanoparticles can also be used as waveguiding modalities, since energy is transferred via near-field coupling between adjacent particles in linear chains. The chapter closes with a description of emerging efforts to combat attenuation via optical gain media as waveguide hosts.

7.1 Planar Elements for Surface Plasmon Polariton Propagation

The propagation direction of SPPs at the interface of a metal film and a dielectric superstrate (air or dielectric) can be controlled via scattering of the propagating, two-dimensional waves at locally created defects in the otherwise planar film. The scatterers can be introduced in the form of surface undulations such as nanoscale particle-like structures, or by the milling of holes into the film. Their controlled positioning enables the generation of functional elements such as Bragg mirrors for reflecting SPPs [Ditlbacher et al., 2002b], or focusing elements for increasing lateral confinement [Yin et al., 2005, Liu et al., 2005]. This way, a planar two-dimensional photonic infrastructure for the guiding of SPPs can be created.

A simple and compelling example of control over SPP propagation via scattering from height modulations was demonstrated by Ditlbacher and coworkers [Ditlbacher et al., 2002b]. Using electron beam lithography and chemical vapor deposition, silica nanostructures such as particles and wires of 70 nm height were deposited on a silica substrate, and the height-modulated film subsequently coated with a 70 nm thick silver film (Figure 7.1). In order to excite SPPs, the method of phase-matching via scattering of the excitation beam (in this case a Ti:sapphire laser beam with $\lambda_0 = 750$ nm) at a nanowire-shaped defect was used (see chapter 3). The SPP propagation pathway was monitored by coating the film with a polymer layer containing fluorescent dyes (see chapter 4). This also enabled an estimate of the 1/e propagation distance of the SPPs at the silver/polymer film, here of the order of 10 μm.

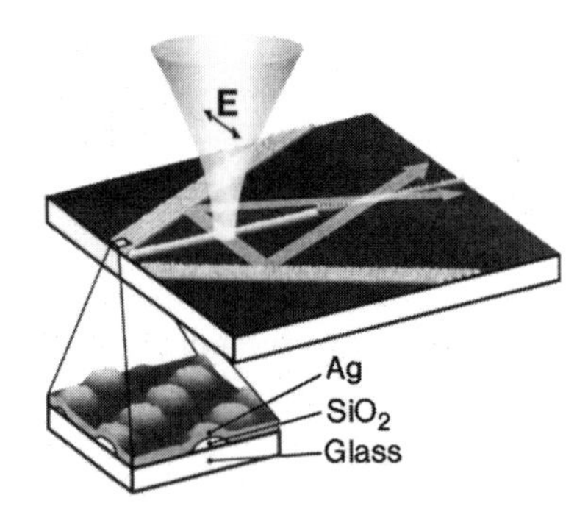

Figure 7.1. Routing of SPPs on a planar silver film using surface modulations. A laser beam focused on a nanowire or nanoparticle defect for phase-matching acts as a local source for SPPs. The micrograph shows a Bragg reflector consisting of lines of regularly spaced, particle-like undulations (Fig. 7.2). Reprinted with permission from [Ditlbacher et al., 2002b]. Copyright 2002, American Institute of Physics.

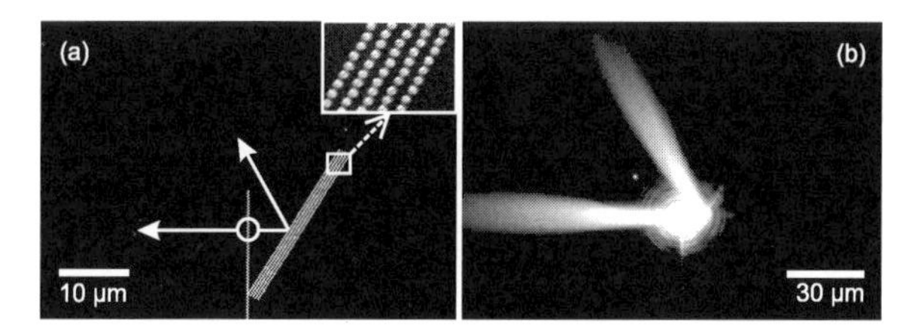

Figure 7.2. (a) SEM image of a SPP Bragg reflector consisting of ordered particle arrays on a metal film substrate. (b) SPP propagation imaged via monitoring of the emission of a fluorescent superstrate. Reprinted with permission from [Ditlbacher et al., 2002b]. Copyright 2002, American Institute of Physics.

Fig. 7.2 shows a Bragg reflector based on this principle, consisting of parallel lines of particles with diameter 140 nm. An interline spacing of 350 nm (Fig. 7.2a) fulfilled the Bragg condition for SPPs impinging at a 60° angle on the array and thus lead to the specular reflection of the SPP wave (fluorescent image in Fig. 7.2b). In this case, the reflection coefficient of a Bragg mirror consisting of 5 lines was estimated to be 90%, with the remaining fraction being scattered out of the plane into radiation. This proof-of-principle study suggests that planar passive optical elements for the routing of SPP propagation can be fabricated in an easy manner. We will show in the next section that the lateral extent of SPPs can be controlled by extending the Bragg mirror concept to create surface plasmon photonic cyrstals exhibiting band gaps for propagation in desired frequency regions.

Another approach for controlling SPP propagation at a single metal interface is the spatial modification of the SPP dispersion and thus phase velocity via dielectric nanostructures deposited on top of the film [Hohenau et al., 2005b], by analogy to the conventional routing of free-space beams with dielectric components such as lenses. Figure 7.3 shows the calculated dispersion relations of SPPs in a glass/gold/superstrate multilayer system for both the s modes (magnetic fields on the two metal interfaces in phase) and the a modes (magnetic fields at the two metal interfaces out of phase) for varying dielectric constants ε_3 of the superstrate. It is evident that an increase in ε_3 leads to an increase in SPP wave vector, as discussed in chapter 2. This implies that the phase velocity of the propagating waves can be locally decreased by introducing dielectric structures on top of the metal film. By adjusting the geometric shape of the dielectric perturbations and thus the regions of reduced phase velocity, it is therefore possible to fabricate optical components such as lenses and waveguides for SPP propagation, albeit with increased attenuation due to the closer confinement of the mode to the metal surface.

Figure 7.4 demonstrates that via this concept, the focusing (top row) and refraction/reflection (bottom row) of SPPs can be achieved using cylindrical- or

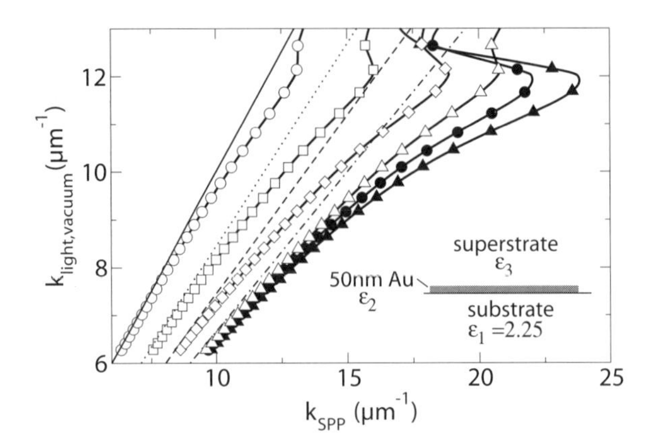

Figure 7.3. Calculated SPP dispersion relations for a glass/gold/superstrate three-layer system for both the *s* mode (open symbols) and the *a* mode (filled symbols). Increasing the dielectric constant ε_3 of the superstrate from $\varepsilon_3 = 1$ (circles) to $\varepsilon_3 = 2.25$ (triangles) leads to an increase in propagation constant and thus a decrease in phase velocity of the SPP. For $\varepsilon_1 = \varepsilon_3$, these two modes would evolve into the symmetric (s) or asymmetric (a) mode. Reprinted with permission from [Hohenau et al., 2005b]. Copyright 2005, Optical Society of America.

triangular-shaped particles, by direct analogy to the three-dimensional optical elements of conventional free-space optics.

In their study, Hohenau and co-workers excited the SPPs with an immersion oil objective and observed SPP propagation via monitoring of leakage radiation (Fig. 7.4 a, b, d, e) and near-field optical microscopy (Fig. 7.4 c, f). The

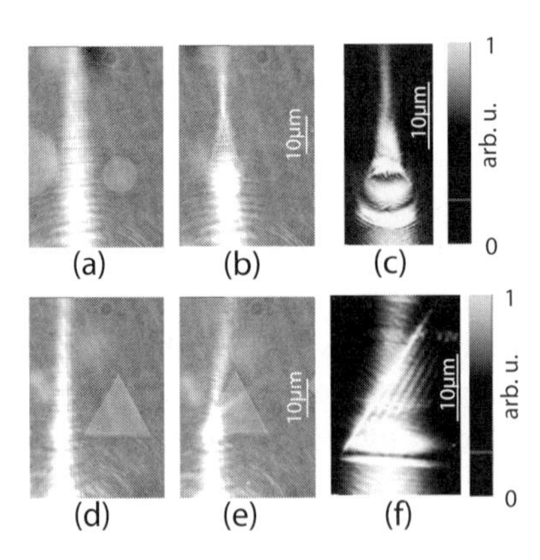

Figure 7.4. Focusing (top row) or reflection and refraction (bottom row) of SPPs via a cylindrical or triangular 40 nm thick dielectric structure deposited on top of a gold film. Images of the leakage radiation (a, b, d, e) and of the optical near field (e, f) clearly show the modification of SPP propagation for SPPs impinging on the dielectric structures (b, c, e, f). Reprinted with permission from [Hohenau et al., 2005b]. Copyright 2005, Optical Society of America.

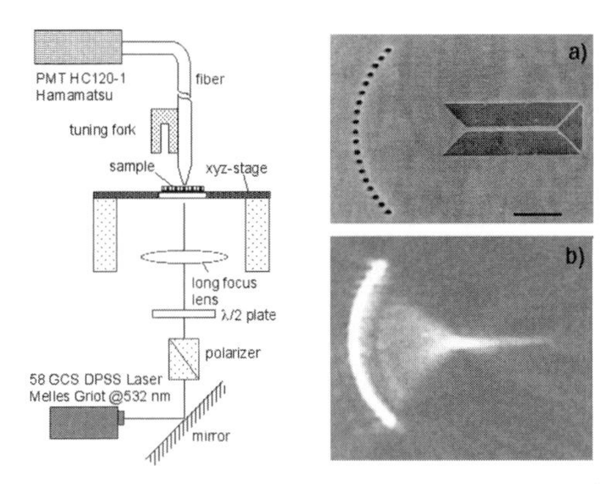

Figure 7.5. Experimental setup of the excitation and near-field imaging for SPP focusing on a holey metal film (left). (a) SEM and (b) near-field optical image of the nanohole focusing array which couples the launched SPPs into a 250 nm wide silver stripe guide. Reprinted with permission from [Yin et al., 2005]. Copyright 2005, American Chemical Society.

same concept should also allow for the creation of the SPP analogue of dielectric waveguides via the creation of one-dimensional regions of reduced phase velocity, which was experimentally confirmed using one-dimensional polymer nanostructures on a gold layer [Smolyaninov et al., 2005].

We will conclude this section by presenting two recent studies of focusing using holes and grooves directly milled into the metallic film sustaining the SPPs. Figure 7.5 shows how constructive interference between SPPs launched locally using illumination of nineteen 200 nm holes arranged on a quarter circle of radius 5 μm in a 50 nm silver film gives rise to a tight focus spot in the center of the circle [Yin et al., 2005]. As an application, Yin et al. used their focusing element for coupling SPPs into a 250 nm wide stripe waveguide (see images a and b in Fig. 7.5).

Excitation and subsequent focusing of SPPs can also be achieved using circular or elliptical sub-wavelength slits milled into a metallic film [Liu et al., 2005]. In this case, the edge of the circular slit acts as a point source for SPPs upon illumination of the slit structure, in regions where the exciting electric field is polarized perpendicular to the slit, and the generated SPPs will be launched and focused towards the center of the circle. The non-resonant nature of this process makes this scheme suitable for focusing SPPs excited at different frequencies throughout the visible spectrum, albeit with low efficiency. Fig. 7.6a shows SPP focusing using a circular slit structure of radius 14 μm and width 280 nm, milled into a 150 nm thick silver film. The near-field pattern for excitation with linearly polarized light was recorded using near-field optical microscopy. As apparent, only two opposite regions of the circle, where the electric field is polarized perpendicular to the slit, act as SPP sources. Illumination with unpolarized light however leads to SPP generation throughout the

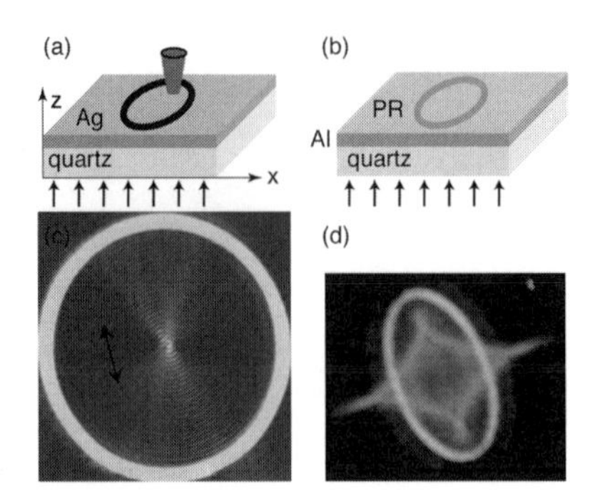

Figure 7.6. Generation and focusing of SPPs via illumination of circular or elliptic slits milled into a metallic film. The SPP intensity is monitored via near-field microscopy (a,c) or the exposure of a photoresist superstrate (b, d). Reprinted with permission from [Liu et al., 2005] Copyright 2005, American Chemical Society.

circumference, shown in Fig. 7.6 for an ellipse of axes 4 μm and 2 μm cut into a 70 nm thick aluminum film. In this case, the near-field pattern was recorded via the exposure of a photoresist layer.

It can be anticipated that the combination of functional elements such as the ones discussed in this section will enable planar photonic circuits working at optical or near-infrared frequencies, with propagation distances below 100 μm.

7.2 Surface Plasmon Polariton Band Gap Structures

The concept of constructively reflecting SPPs on a metal film via Bragg reflectors created using periodically arranged metallic nanoparticles presented in Figs. 7.1 and 7.2 can be extended to the creation of band gaps for SPP propagation using regular metal nanoparticle lattices deposited on a metal film. Bozhevolnyi and co-workers demonstrated that a triangular lattice of gold dots on a thin gold film establishes a band gap for SPP propagation [Bozhevolnyi et al., 2001]. An example of such a structure is shown in Fig. 7.7 for a triangular lattice of gold scatters fabricated on a 40 nm thin gold film. In this case, the lattice constant was chosen to be 900 nm, and the individual scatters are 378 nm wide and 100 nm high, resulting in the formation of a band gap in the telecommunication window (wavelengths around $\lambda = 1.5$ μm) [Marquart et al., 2005]. The penetration of SPPs (excited via prism coupling on the flat parts of the film) incident on this structure can be monitored using near-field optical microscopy, and examples of near-field images obtained at two different wavelengths are shown in panels (b) and (c). This way, the band gap for SPP propagation can be determined for a given direction of the incident SPPs by determining the penetration distance of the surface waves into the lattice structure.

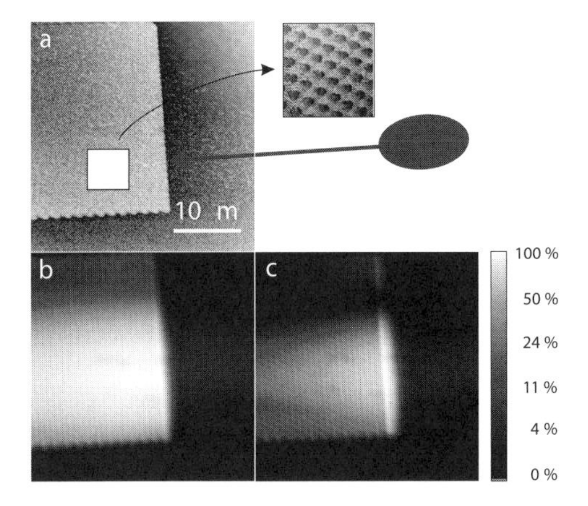

Figure 7.7. Topography (a) and near-field optical images (b,c) of a $35 \times 35\ \mu m^2$ SPP band gap structure consisting of a 900 nm triangular lattice of 378 nm wide and 100 nm high gold dots on a 40 nm thick gold film. SPPs excited via prism coupling of radiation of wavelength 1550 nm (b) or 1600 nm (c) propagate from the right into the lattice structure in the ΓK direction, and are strongly attenuated if their frequency is inside the band gap (c). Reprinted with permission from [Marquart et al., 2005]. Copyright 2005, Optical Society of America.

An application of this concept in waveguiding is obvious: by creating micron-wide line defects where the triangular lattice of scatters is locally removed, SPPs can be laterally confined in channel waveguides, akin to well established concepts in planar dielectric photonic crystals. Figure 7.8 shows a near-field optical image of SPPs excited at $\lambda_0 = 1550$ nm via prism coupling, guided within a channel defect waveguide in a triangular lattice of gold dots separated by a period 950 nm. Note that in this case parts of the guided SPPs

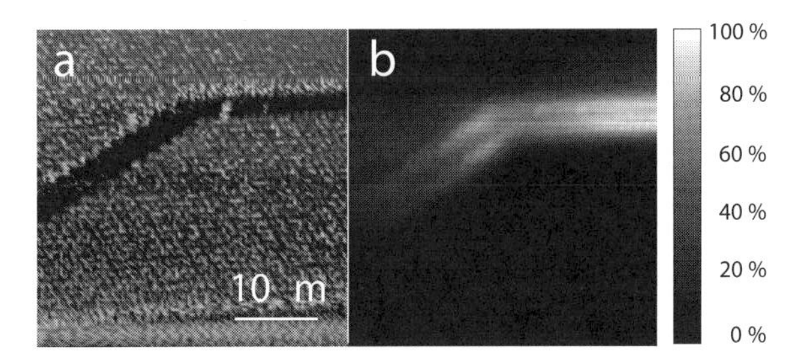

Figure 7.8. Topographical (a) and near-field optical (b) image of a channel defect waveguide in a triangular lattice of period 950 nm consisting of 438 nm wide and 80 nm high gold scatters on an gold film. A SPP excited at $\lambda_0 = 1515$ nm incident from the right propagates through the channel. Reprinted with permission from [Marquart et al., 2005]. Copyright 2005, Optical Society of America.

inside the channel leak into the surrounding lattice at the channel bend, since the band gaps for different directions in the irreducible Brillouin zone do not overlap. Because in waveguides based upon this principle the lateral confinement given by the channel width is of the order of the vacuum wavelength, the $1/e$ decay length of the guided SPP waves is comparable to that of the unmodulated, flat interface of the respective metal/dielectric system.

7.3 Surface Plasmon Polariton Propagation Along Metal Stripes

We now move on to multilayer structures and their use in waveguiding applications. In this section we present a particularly simple concept of a waveguide for SPPs with controlled lateral confinement. It is based upon the insulator/metal/insulator multilayer system described in chapter 2 and consists of a thin metal stripe sandwiched between two thick dielectric cladding layers (Fig. 2.5). We have seen that for a sufficiently thin metallic core layer of thickness t, interactions between SPPs on the bottom and top interfaces lead to the occurrence of coupled modes. For a symmetric system with equal dielectric sub- and superstrate, the modes are of well-defined symmetries, and the odd mode (defined as in chapter 2) displays the intriguing property of dramatically decreased attenuation with a reduction in metal thickness. As described previously, this is due to decreasing confinement of the mode as it evolves into the TEM mode propagating in the homogeneous background dielectric for $t \to 0$.

Whereas our treatment in chapter 2 dealt exclusively with multilayer structures of infinite width w, here we will present a number of studies of coupled SPP modes guided along metallic stripes of finite width. We will restrict our discussions to waveguides of cross sections with $w/t \gg 1$, where only the vertical dimension t is sub-wavelength (see sketch in Fig. 7.9). Guiding along nanowires where additionally $w < \lambda_0$ will be discussed in the next section. Before presenting the case of a metallic stripe on a dielectric substrate with air as the superstrate, we will first address the important case of metallic stripes embedded in a homogeneous dielectric environment. We have already seen in chapter 2 that a long-ranging SPP mode is supported for infinitely wide structures. This is also true for stripes of appreciable but finite w, which is the

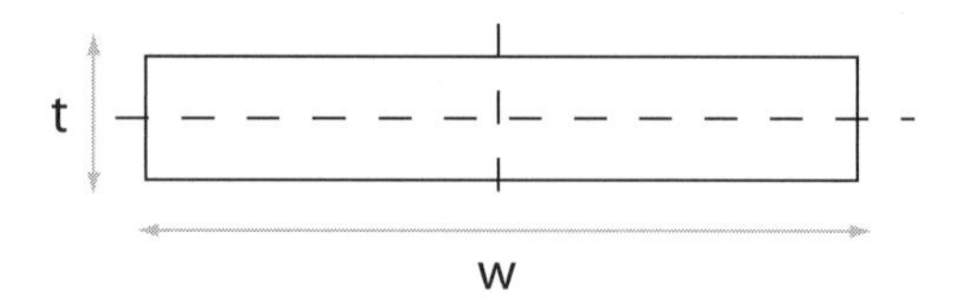

Figure 7.9. Cross section of a metal stripe waveguide of finite width. The dashed lines depict the symmetry planes.

reason that this geometry has received a great amount of attention for practical applications in waveguiding.

Berini presented a theoretical study of bound modes supported by such a thin metal stripe embedded in a homogeneous dielectric host [Berini, 2000]. As well as two fundamental modes of opposite symmetries, which retain much in character of the two coupled modes of the infinite layer system, his study comprehensively analyzes the different higher-order bound modes sustained by this structure. The bound modes are classified by two letters describing the symmetry of the electric field component perpendicular to the long stripe edges with respect to the two symmetry planes of the stripe (dashed lines in Fig. 7.9), and a number denoting how many field nodes are encountered along the stripe width. In this notation, the fundamental bound mode we want to focus on is denoted as ss_b^0. It closely resembles the odd bound mode of the infinitely wide symmetric structure (in this notation called s_b rather than a_b due to different conventions of classifying symmetry either with respect to the component of the electric field perpendicular to the long edges as in this case, or with respect to the component parallel to the direction of propagation as in the description of chapter 2).

The calculated dispersion of the first four modes of a silver stripe of width $w = 1\ \mu$m with thickness t for a symmetric host material with $\varepsilon = 4$ is shown in Fig. 7.10 for excitation at a vacuum wavelength $\lambda_0 = 633$ nm, together with the results for the two modes s_b and a_b sustained by the infinitely wide multilayer geometry. The evolution of the real part of the propagation constant β (normalized to the free space value β_0) is shown in Fig. 7.10a, while Fig. 7.10b shows the imaginary part of β, representing the attenuation suffered by the traveling coupled SPP waves. We will not describe the evolution of the modes in detail, but want to draw attention on the fundamental ss_b^0 mode, which is seen to evolve similarly to the (long-ranging) s_b mode of the infinite structure. This mode does not show a cut-off thickness, and its attenuation dramatically decreases over many orders of magnitude with decreasing stripe thickness. In analogy to the infinitely wide slab [Sarid, 1981], this mode is called the *long-ranging* SPP mode of the stripe.

As we can expect after the discussion of the long-ranging mode for the infinitely wide system in chapter 2, the decrease in attenuation with film thickness is accompanied by an equally dramatic loss in confinement, as the mode evolves into the TEM mode of the host for vanishing stripe thickness: the mode extends over many wavelengths into the dielectric host medium as its confinement (defined by the fraction of the power flowing through the stripe itself to the total power in the mode) decreases with thickness. This loss in confinement seems to be exacerbated for stripes with widths below λ_0. From the point of view of strong confinement and high integration density, insulator/metal/insulator waveguides are thus clearly not the favorable geometry

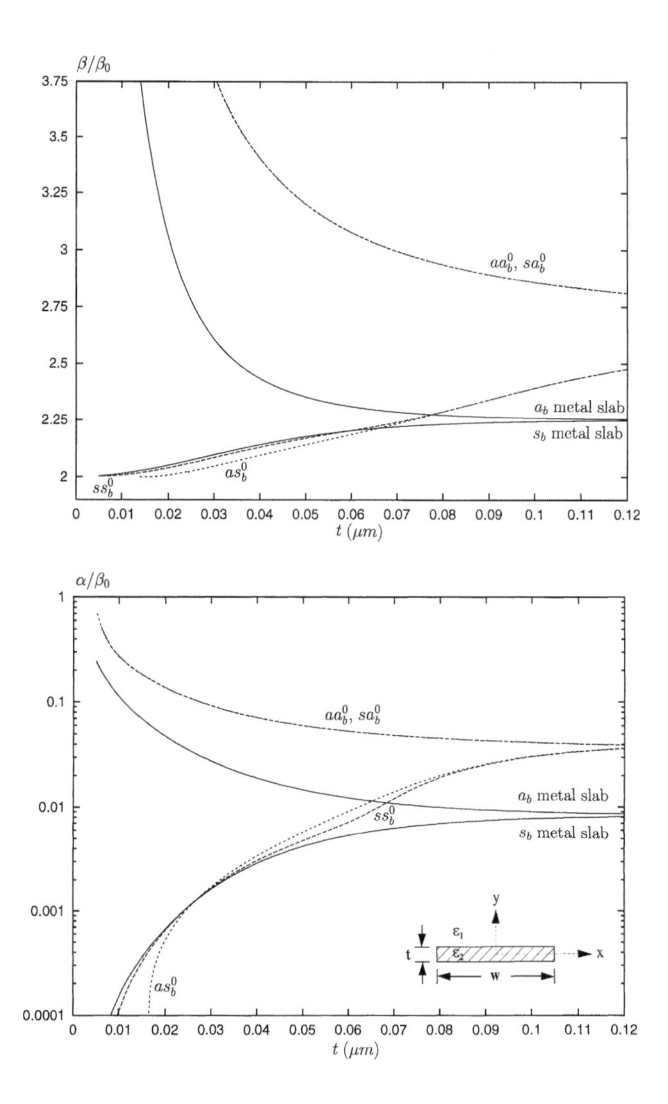

Figure 7.10. Evolution of the propagation constant for the first four modes of a 1 μm wide silver stripe embedded in a homogeneous medium with dielectric constant $\varepsilon = 4$ for excitation at a vacuum wavelength $\lambda_0 = 633$ nm. Also shown are the symmetric and antisymmetric modes of the infinitely wide interface (denoted as metal slab). (a) normalized phase constant. (b) normalized attenuation constant. Reprinted with permission from [Berini, 1999]. Copyright 1999, Optical Society of America.

of choice [Zia et al., 2005c]. We refer the reader to the original publication [Berini, 2000] presenting a detailed analysis of the evolution of the long-ranging mode with w, the dielectric constant of the host, and excitation wavelength. Also, in a follow-up on his original work, Berini analyzed stripes embedded in an asymmetric environment, demonstrating that the long-ranging mode is absent in this case, due to the phase mismatch between the SPPs at the two different metal/insulator interfaces [Berini, 2001].

The properties of the long-ranging mode serve us as a good demonstration of the general principle of the trade-off between localization and loss occurring in plasmon waveguides, which we will encounter throughout this chapter. Since tight field localization to the metal interfaces necessarily implies that a significant amount of the total mode energy resides inside the metal itself, the propagation loss increases due to Ohmic heating. Thus, as we will see, guiding of electromagnetic energy with sub-wavelength mode confinement will imply micron or even sub-micron propagation lengths. The long-ranging SPP modes of metal stripes on the other hand can show 1/e attenuation lengths approaching 1 cm in the near-infrared, due to the low confinement for a film thickness on the order of 20 nm.

From an application point of view, the long-ranging mode exhibits the additional desirable property that its spatial field profile exhibits a Gaussian-like

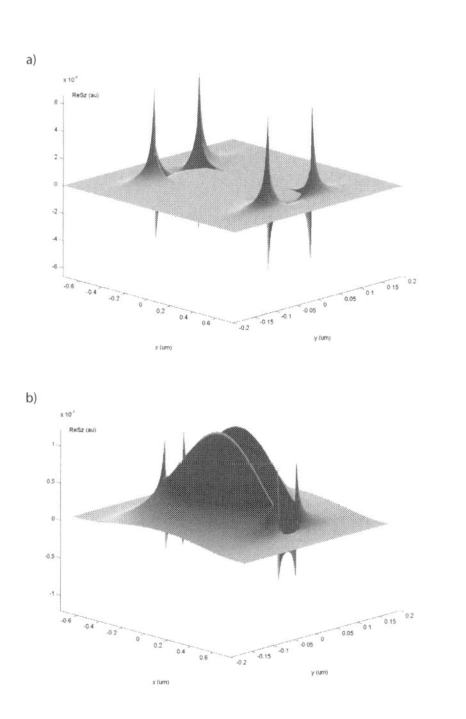

Figure 7.11. Mode profile of the real part of the Poynting vector for the long-ranging ss_b^0 mode at $\lambda_0 = 633$ nm of a 100 nm (a) or 40 nm (b) thick and 1 μm wide silver stripe, showing the Gaussian-like mode shape for small film thickness. Reprinted with permission from [Berini, 1999]. Copyright 1999, Optical Society of America.

lateral distribution for small thickness t [Berini, 2000]. Fig. 7.11 shows the spatial distribution of the real part of the Poynting vector for a 100 nm (a) and 40 nm (b) thin stripe. For the thick stripe, the energy is mostly guided along the edges (Fig. 7.11a), while for the thin stripe, the Gaussian shape (Fig. 7.11b) enables efficient end-fire coupling via spatial mode-matching.

The first experimental demonstration of the long-ranging mode employed a $t = 20$ nm thick and $w = 8$ μm wide gold stripe embedded in glass, and guiding over multiple millimeters was demonstrated [Charbonneau et al., 2000]. More quantitative studies of its propagation characteristics followed. For 10 nm thick stripes of similar widths embedded in a polymer host, a propagation loss of only $6 - 8$ dB/cm at $\lambda_0 = 1550$ nm has been experimentally confirmed [Nikolajsen et al., 2004a]. Also, long-range SPP propagation along sub-wavelength nanowires has been observed [Leosson et al., 2006], albeit with the mode extending appreciably into the homogeneous dielectric background as expected.

The long propagation distances and micron-sized widths (allowing lateral structuring) of stripe waveguides have already enabled the demonstration of useful optical elements such as bends and couplers [Charbonneau et al., 2005], Bragg mirrors engraved directly on the waveguide [Jette-Charbonneau et al., 2005], and integrated power monitors based on direct detection of Ohmic heat generation [Bozhevolnyi et al., 2005a]. Also, active switches and modulators operating on the same thermal principle have been demonstrated [Nikolajsen et al., 2004b]. It remains to be seen at which point these waveguides will find their first commercial applications.

We will now discuss a second important stripe waveguide geometry, namely that of a metal stripe layer on a dielectric substrate surrounded by air. Due to the large dielectric asymmetry between the substrate and the superstrate, in this geometry the long-ranging mode is absent. A comprehensive survey of the propagation lengths exhibited by such stripes has been performed by Lamprecht and co-workers, who studied SPP propagation along 70 nm thick gold and silver stripes with widths $1 \leq w \leq 54$ μm [Lamprecht et al., 2001]. SPPs on the top metal/air interface were excited using a prism coupling arrangement with a shielding layer to prevent direct excitations along the length of the stripe (Fig. 7.12), and SPP propagation was monitored via the collection of the light scattered via surface roughness. A dramatic decrease in propagation length with decreasing stripe width was observed as the width of the stripe became comparable with the wavelength of excitation (Fig. 7.13, data points).

Apart from the significantly smaller propagation length in comparison to the SPP modes sustained by the stripes embedded in a homogeneous medium discussed above, it is important to note that the modes excited on the metal/air interface in stripes using prism coupling are inherently *leaky modes*, as discussed in chapter 3. The propagating modes are not only attenuated due to absorption, but also due to re-radiation into the higher-index substrate. End-

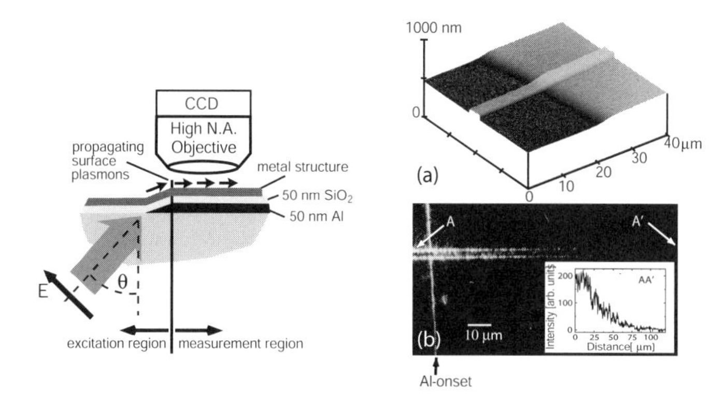

Figure 7.12. Prism coupling setup for the excitation of leaky SPPs propagating on thin metal stripes (left). The aluminum screen shields the stripe from direct excitation along its length. (a) AFM image of a 3 μm wide stripe. (b) Scattered light image showing the propagating SPP excited at $\lambda_0 = 633$ nm. Reprinted with permission from [Lamprecht et al., 2001]. Copyright 2001, American Institute of Physics.

fire excitation of stripe modes in a homogeneous medium on the other hand can excite the truly bound modes of the system.

Using a full-vectorial, magnetic finite-difference method, Zia and co-workers solved for the fundamental and higher-order leaky modes sustained by metallic stripes that are excited in prism coupling experiments [Zia et al., 2005b]. As shown in Fig. 7.13, the computed propagation lengths of the lowest-order

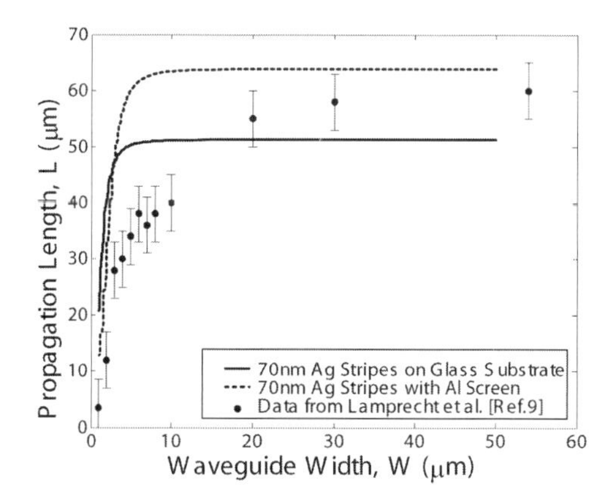

Figure 7.13. Comparison of experimental results (data points) for the SPP propagation length of thin silver stripes [Lamprecht et al., 2001] with numerical modeling of lowest-order, quasi-TM leaky modes (curves). Reprinted with permission from [Zia et al., 2005b]. Copyright 2005 by the American Physical Society.

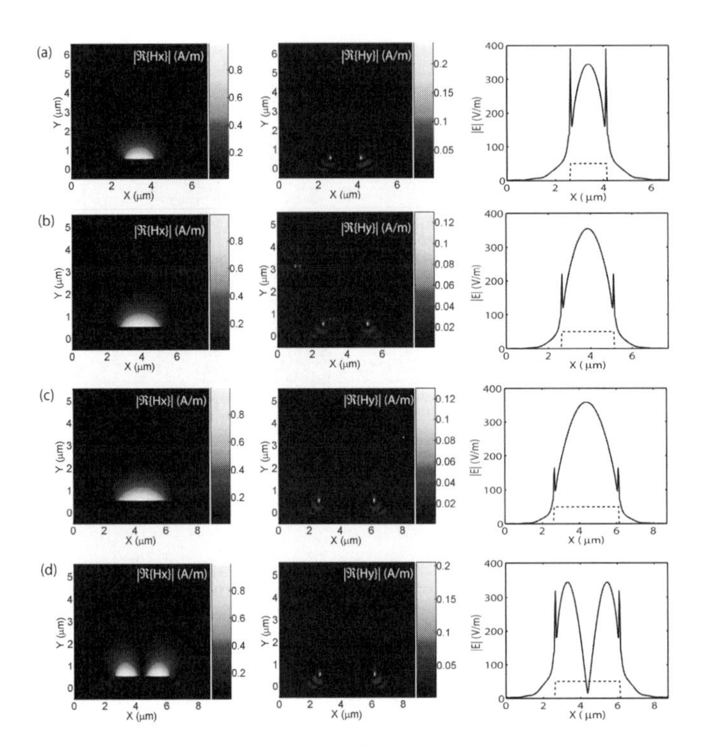

Figure 7.14. Transverse magnetic field profiles (first and second column) and electric field intensities (third column) for leaky, quasi-TM SPP modes of gold stripe waveguides (t = 55 nm, λ_0 = 800 nm) for (a) w = 1.5 μm (sole, lowest-order mode), (b) w = 2.5 μm (sole, lowest-order mode), (c) w = 3.5 μm (lowest-order mode) and (d) w = 3.5 μm (second-order mode). Reprinted with permission from [Zia et al., 2005b]. Copyright 2005 by the American Physical Society.

quasi-TM (i.e., the mode that is of TM polarization in the symmetry plane) leaky mode are in good agreement with the experimental results obtained by Lamprecht and colleagues [Lamprecht et al., 2001] when the shielding layer is taken into account. The calculated mode profile of the fundamental and first higher-order quasi-TM leaky modes for gold stripes of different widths are depicted in Fig. 7.14, together with cross cuts of the near-field intensity profile above the stripes. The numerically determined intensity distribution compares well with experimental near-field optical investigations using prism coupling and collecting the near field using an apertured fiber tip [Weeber et al., 2003]. As an example, Fig. 7.15 shows topographical images and the collected near field above gold stripes of height 55 nm and widths 3.5 μm or 2.5 μm, clearly visualizing the propagating SPP waves. Transverse cuts through the near-field intensity distribution (Fig. 7.16) are similar to the calculated distribution of the electric field (Fig. 7.14, third column). We note that since the apertured tips

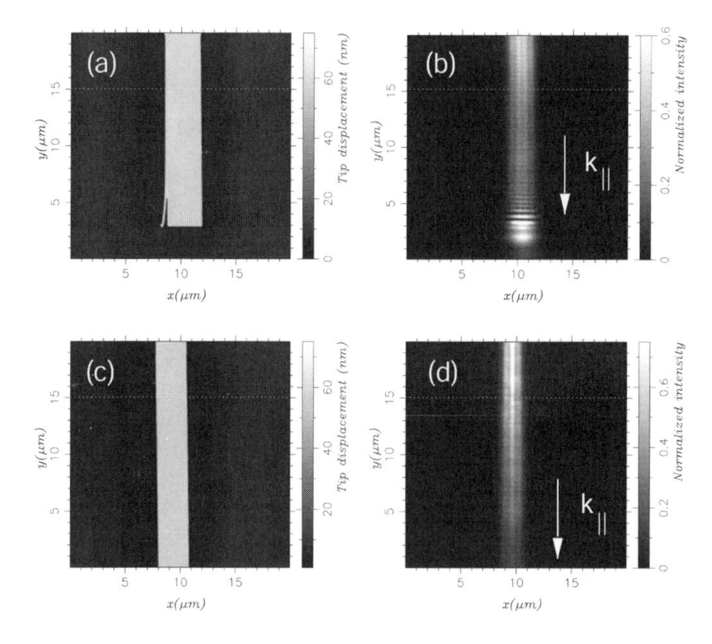

Figure 7.15. AFM (a and c) and near-field optical (b and d) images of gold stripes of height 55 nm and width $w = 3.5$ μm (a and b) or $w = 2.5$ μm (c and d). Reprinted with permission from [Weeber et al., 2003]. Copyright 2003 by the American Physical Society.

used for collecting the fields in this study were coated with a thin chromium layer (exhibiting only negligible conductivity at the excitation frequency), the collected near-field images are indeed expected to follow the distribution of the electric field plotted in the third column of Fig. 7.14.

Apart from explaining the observed near-field distribution and therefore the mode structure of leaky modes excited via prism coupling, another significant outcome of this numerical study is the existence of a lower bound to the stripe width below which no propagating leaky modes exist for this geometry. The numerical studies are further corroborated by an intuitive dielectric waveguide model of SPP stripe waveguides [Zia et al., 2005a], which shows that the well-established treatment of dielectric waveguides [Saleh and Teich, 1991] can be applied to SPP waveguides if the effective index n_{eff} is calculated via the SPP dispersion as $n_{\text{eff}} = \frac{\beta}{k_0}$. This suggests that the transverse dimensions of SPP stripe waveguides have to obey a *diffraction limit* $\Delta x \geq \frac{\lambda_0}{2n_{\text{eff}}}$, limiting the amount of transverse confinement and thus the integration density of such waveguides. However, experimental evidence for SPP propagation with large confinement along nanowires has been obtained by a number of groups (see next section), so that further clarification of the constraints on transverse confinement of stripe waveguides is needed.

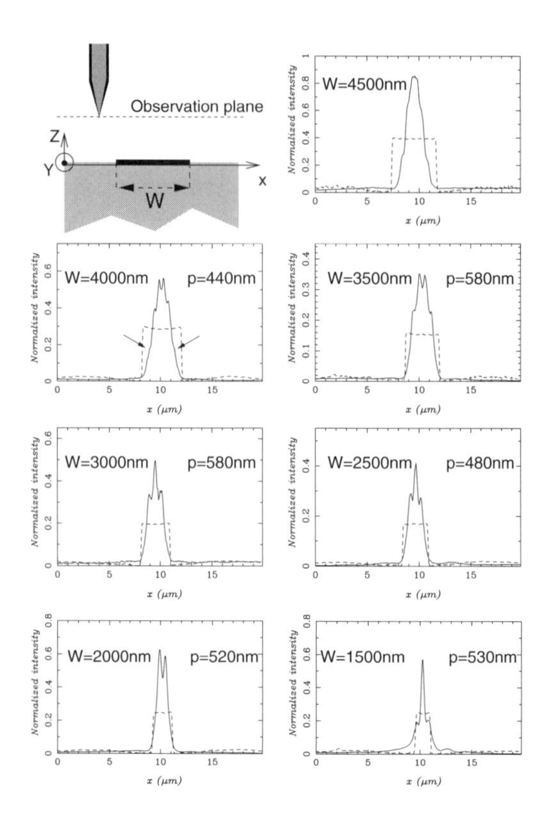

Figure 7.16. Cross-cuts through the near-field intensity of various stripes of width w (see Fig. 7.15). p denotes the distance between the peaks. Compare with the calculated profiles in the third column of Fig. 7.13. Reprinted with permission from [Weeber et al., 2003]. Copyright 2003 by the American Physical Society.

As with the long-range SPP waveguides discussed above, first demonstrations of functional elements placed directly on stripe waveguides are emerging, such as Bragg mirrors [Weeber et al., 2004] or triangular shaped terminations for modest SPP field focusing [Weeber et al., 2001]. Integration with conventional silicon waveguides has also been demonstrated [Hochberg et al., 1985], and the use of SPP stripe waveguides to guide energy around sharp bends coupled to to Si waveguides has been suggested.

7.4 Metal Nanowires and Conical Tapers for High-Confinement Guiding and Focusing

The fact that metal waveguides of a cross section substantially below the square of the wavelength λ of the guided radiation can exhibit transverse mode confinement below the diffraction limit in the surrounding dielectric can be easily derived using the uncertainty relation between the transverse components of the wave vector and the corresponding transverse spatial coordinates [Takahara et al., 1997]. To see this, we recall the simple argument why the mode size of waves guided along the core of a dielectric waveguide is limited by diffraction. For propagation along the z-direction, the relationship between propagation constant β, the transverse components of the wave vector k_x, k_y and the frequency ω of the guided radiation is given by

$$\beta^2 + k_x^2 + k_y^2 = \varepsilon_{\text{core}} \frac{\omega^2}{c^2}. \tag{7.1}$$

Since in a dielectric waveguide $\varepsilon_{\text{core}} > 0$ and k_x, k_y are real, (7.1) implies that $\beta, k_x, k_y \leq \sqrt{\varepsilon_{\text{core}}}\omega/c = 2\pi n_{\text{core}}/\lambda_0$. According to the uncertainty relation between wave vector and spatial coordinates, the mode size of such *three-dimensional* optical waves is thus limited by the effective wavelength in the core medium:

$$d_x, d_y \geq \frac{\lambda_0}{2n_{\text{core}}} \tag{7.2}$$

However, if the guiding medium in the core is of metallic character, then $\varepsilon_{\text{core}} < 0$ (ignoring for simplicity attenuation). In order for (7.1) to be fulfilled, either one or both of the transverse wave vector components k_x, k_y must be imaginary - the guided waves are two- or one-dimensional, respectively. In this case, relation (7.2) does not apply, and the mode size can be substantially below the diffraction limit of the surrounding dielectric cladding. As our discussion in chapter 2 has shown, we can expect that also the *effective mode area*, taking into account the energy of the mode in the metal itself, should be below the diffraction limit. We point out however that metallic guiding structures of sub-wavelength cross section do not necessarily support such highly confined modes, as was pointed out in our discussion of the long-ranging SPP modes earlier on.

Studies of metal nanowire waveguides - essentially the same type of metal stripe waveguides on a dielectric substrate discussed above, but with a sub-wavelength width - have indeed provided evidence for leaky mode propagation of SPPs excited in prism-coupling geometries using both conventional [Dickson and Lyon, 2000] and collection-mode near-field optical microscopy [Krenn et al., 2002] to image the guided surface waves. To illustrate the guiding capabilities of such structures, Fig. 7.17 shows the topography (a) and a near-field optical image (b) of a 20 μm long gold nanowire with $w = 200$ μm and $t = 50$ nm [Krenn et al., 2002]. A leaky SPP mode was excited on the wire at $\lambda_0 = 800$ nm using the same prism coupling launch-pad technique depicted in Fig. 7.12. The collected near-field intensity above the wire is indicative

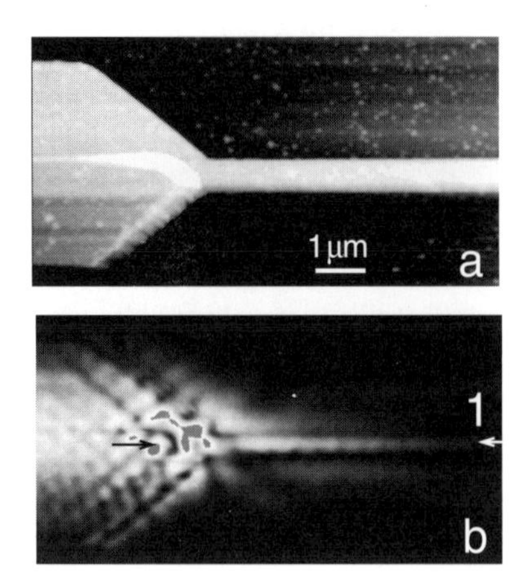

Figure 7.17. Topography (a) and optical near-field intensity (b) of a 20 μm long gold nanowire of width w = 200 nm excited at λ_0 = 800 nm. The arrows mark the position of data-cut 1 shown in Fig. 7.18. Reproduced with permission from [Krenn et al., 2002]. Copyright 2002, Institute of Physics.

of electromagnetic energy transfered along the wire axis. Fig. 7.18a shows a cross-cut through the near-field intensity along the wire axis (solid line), which can be fitted by an exponential decay with attenuation constant $L = 2.5\ \mu$m (dashed line). The deduced SPP propagation length is significantly shorter than that of stripes with widths in excess of a couple of micrometers, in line with the steep decline in propagation length observed in Fig. 7.13 [Lamprecht et al., 2001]. If the length of the wire is shortened, an oscillation in near-field intensity is established, indicative of standing waves due to reflection of the SPPs at the end-facet (Fig. 7.18a, inset). In order to judge the transverse confinement, Fig. 7.18b shows a cross-cut through the optical near-field intensity perpendicular to the wire axis. As can be seen, the fields are essentially localized to the physical extent of the wire.

It has to be pointed out that the apparent observation of SPP guiding in the prism-excited leaky mode is in contradiction to the theoretical work by Zia and co-workers (discussed in the previous section) that claimed that the fundamental leaky mode sustained by the stripe is cut off below a certain width. Since as pointed out above their study showed remarkable agreement with near-field optical investigations of stripes with $w \geq 1\ \mu$m, the nature of the mode observed in the study by Krenn and colleagues requires further theoretical clarification.

In addition to the excitation of a leaky mode along a nanowire, a truly bound mode outside the light cone of the substrate can be excited by changing the excitation condition from prism-coupling to coupling using a high-NA objective. Ditlbacher and co-workers have used this technique to excite a bound SPP propagating along a 18.6 μm long silver wire with w = 120 nm [Ditlbacher

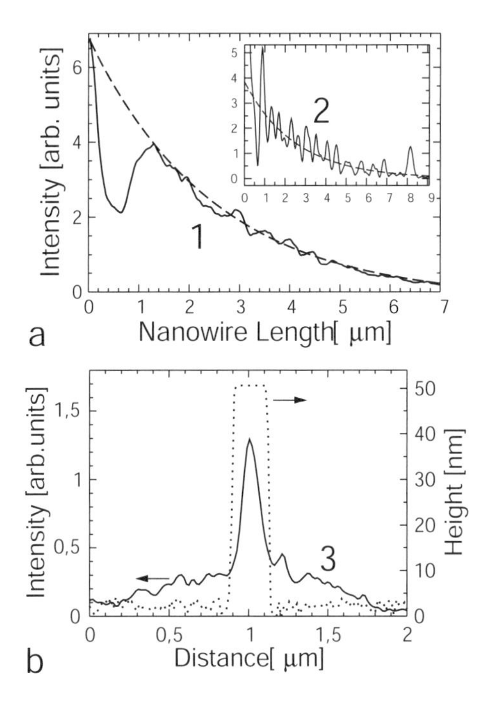

Figure 7.18. (a) Curve 1: optical near-field intensity along the axis of the 20 μm long nanowire of Fig. 7.17 (solid line) and exponential fit with decay constant $L = 2.5$ μm (dashed line). Curve 2: ditto for a 8 μm long wire, showing an interference pattern due to reflections. (b) Cross-cut of the optical near-field intensity (solid line) perpendicular to the wire axis and topography profile as determined by SEM (dotted line). Reproduced with permission from [Krenn et al., 2002]. Copyright 2002, Institute of Physics.

et al., 2005]. Using far- and near-field optical microscopy, a comparatively large SPP propagation length $L \approx 10$ μm has been confirmed. This hugely increased propagation length compared to the initial nanowire study can possibly be attributed to the fact that the mode excited using focused illumination is a bound mode, thus not suffering losses due to leakage radiation into the supporting substrate. Additionally, the nanowires under study were prepared using a chemical synthesis method instead of electron beam lithography, resulting in a highly crystalline structure, further decreasing losses. Reflection of the SPPs at the end facet of the nanowire lead to a resonant structure under white light illumination, with the short nanowire acting as a SPP resonant cavity with sub-wavelength transverse cross section. The fact that nanostructures synthesized by chemical means show an improvement in guiding performance seems highly promising.

These encouraging results in terms of transverse mode confinement with yet appreciable propagation lengths in excess of 1 μm suggest that metal nanowires

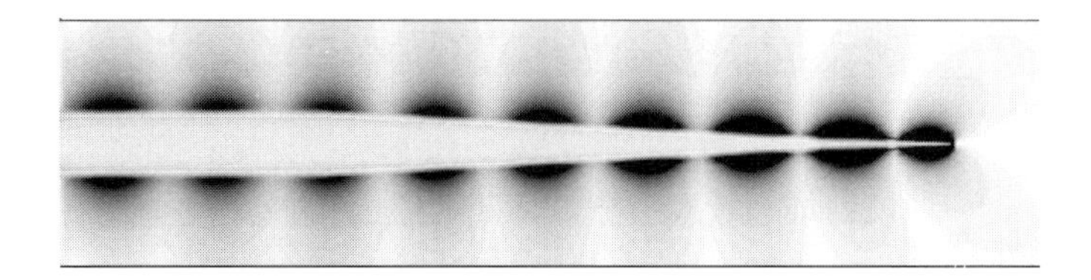

Figure 7.19. Distribution of the electric field around a tapered silica fiber coated with a silver layer of thickness 40 nm. The full taper angle is 6°, and the initial radius of the silica taper is 160 nm. The apex is terminated with a 10 nm semi-sphere. Transfer of energy from the fiber to the plasmon mode and energy concentration is visible ($\lambda_0 = 1.3\ \mu$m).

can be used for creating minituarized photonic circuits for electromagnetic energy transport at visible frequencies [Takahara et al., 1997, Dickson and Lyon, 2000]. It remains to be seen whether this geometry or the metal/insulator/metal gap waveguide geometry discussed in the next section will be more amenable for practical applications.

Before moving on, we want to briefly discuss the possibility of adiabatically increasing the transverse mode confinement along a wire. It can be intuitively expected that the high localization of the optical energy to the surface of a metal nanowire opens up the possibility of further field focusing by creating conically shaped nanowire tapers (Fig. 7.19). Using an analytical boundary problem analysis of the conical geometry of a metal tip, Babadjanyan and coworkers suggested that the shortening of the wavelength as the SPPs propagate along the taper to regions of ever-decreasing diameter enables nanofocusing, with accompanying giant field enhancement at the apex [Babadjanyan et al., 2000]. This was further corroborated using a WKB analysis of the problem, also suggesting that the travel time of SPPs to an infinitely sharp tip should be logarithmically divergent [Stockman, 2004]. A careful analysis of non-local effects on the SPP dispersion occurring in regions with small taper diameter on the order of a few nm close to the apex has further confirmed the focusing properties of such tapers [Ruppin, 2005]. Apart from applications in planar geometries, the experimental realization of such superfocusing structures could potentially be of great use in optical investigation of surfaces in near-field optical microscopy. As an example, Fig. 7.19 shows the electric field distribution of a radially symmetric mode of a nanotaper in cross-cut along its axis, demonstrating the reduction in wavelength and accompanying increased localization and thus field-enhancement as the tip is approached. In this case, the nanotaper consists of a conventional silica fiber taper coated with a thin silver film. Power transfer from the fiber to the plasmon mode takes place, and the energy is then further concentrated as the mode propagates to the apex.

7.5 Localized Modes in Gaps and Grooves

In our discussion of metallic stripes embedded into a homogeneous host, we have only focused on the long-ranging SPP mode with low field localization. Other modes such as the asymmetric sa_b^0 or aa_b^0 offer *sub-wavelength* confinement perpendicular to the interfaces (Fig. 7.10) [Berini, 2000]. Also, the investigations of metallic nanowires presented in the preceding section suggest that such structures allow a transverse mode area smaller than the diffraction limit. An additional and easily amenable structure (both analytically and experimentally) offering sub-wavelength confinement are metal/insulator/metal waveguides, where the mode is confined to the dielectric core in the form of a coupled gap-SPP between the two interfaces. We have analyzed the sub-wavelength energy localization offered by the fundamental mode sustained by this structure in chapter 2, demonstrating that even though upon decreasing gap-size an appreciable fraction of the total mode energy resides inside the metal, increased localization to the interface leads to a high electric field inside the dielectric core, pushing the effective mode length of the one-dimensional system into the deep sub-wavelength region. Therefore, the mode confinement below the diffraction limit of metal/insulator/metal waveguides could enable integrated photonic chips with a high packing density of waveguiding modalities [Zia et al., 2005c].

Two-dimensionally localized modes in SPP gap waveguides have been analyzed analytically both in vertical geometries [Tanaka and Tanaka, 2003] - resembling the discussion in chapter 2 - and in planar analogues [Veronis and Fan, 2005, Pile et al., 2005]. An experimental proof-of-concept realization of the latter gap geometry has further established that end-fire coupling to waveguides with even sub-wavelength slot widths is possible [Pile et al., 2005].

Another simple geometry of SPP gap waveguides are grooves of triangular shape milled into a metal surface. Analytical [Novikov and Maradudin, 2002] and FDTD studies [Pile and Gramotnev, 2004] have suggested that a bound SPP mode exists at the bottom of the groove, offering sub-wavelength mode confinement. Due to the phase mismatch between the SPP modes propagating at the bottom of the groove and the inclined plane boundaries, the mode stays confined at the bottom without spreading laterally upwards. Qualitatively, the dispersion of the mode is similar to that in planar structures [Bozhevolnyi et al., 2005b]. Experimentally, it was shown that 0.6 μm wide and 1 μm deep grooves milled into a gold surface (using a focused ion beam) guide a bound SPP mode in the near-infrared telecommunications window with a propagation length on the order of 100 μm and a mode width of about 1.1 μm [Bozhevolnyi et al., 2005b]. The appreciable propagation length offered by this geometry allows the creation of functional photonic structures. Examples of SPP propagation at $\lambda_0 = 1500$ nm are shown in Figs. 7.20 and 7.21 for a number of functional structures such as waveguide splitters, interferometers and couplers

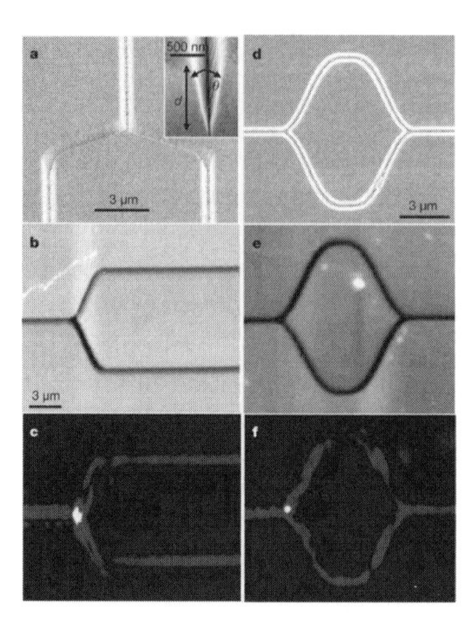

Figure 7.20. SEM (a, d), topographical (b, e) and near-field optical (c, f) images of SPP groove waveguides milled into a metal film. Reprinted by permission from Macmillan Publishers Ltd: Nature [Bozhevolnyi et al., 2006], copyright 2006.

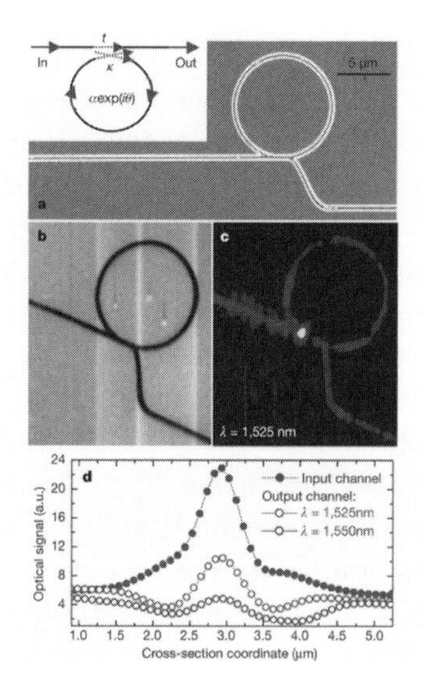

Figure 7.21. SEM (a), topographical (b) and near-field optical images (c) of a channel drop filter based on a V-groove waveguide and a ring resonator. Panel (d) shows normalized cross sections of the input and output channel obtained from (c) for two different wavelengths, demonstrating the extinction ratio on resonance. Reprinted by permission from Macmillan Publishers Ltd: Nature [Bozhevolnyi et al., 2006], copyright 2006.

to ring-waveguides for filtering [Bozhevolnyi et al., 2006]. However, in this study the dimensions of the groove and the guided modes are not appreciably sub-wavelength, explaining the relatively large propagation length compared to those found in nanowires or particle chain waveguides, which will be presented next.

7.6 Metal Nanoparticle Waveguides

Another concept for guiding electromagnetic waves with a transverse confinement below the diffraction limit is based on near-field coupling between closely spaced metallic nanoparticles. As we have seen in chapter 5, a one-dimensional particle array can exhibit coupled modes due to near-field interactions between adjacent nanoparticles. For a center-to-center spacing $d \ll \lambda$, where λ is the wavelength of illumination in the surrounding dielectric, neighboring particles couple via dipolar interactions, with the near-field term scaling as d^{-3} dominating.

Due to the coupling, the nanoparticle chain supports one longitudinal and two transverse modes of propagating polarization waves. The transport of energy along such a chain has been analyzed in a number of approximations, starting with the initial study by Quinten and co-workers based on Mie scattering theory [Quinten et al., 1998]. While this study hinted at the possibility of energy transfer and arrived at estimates of sub-micron energy propagation lengths, subsequent studies concentrated on the dispersion properties. A representation of the particles as point-dipoles allowed the computation of the quasi-static dispersion relation, shown as solid curves in Fig. 7.22 for both longitudinal and transverse polarisation [Brongersma et al., 2000]. The *group velocity* for energy transport, given by the slope of the dispersion curves, is highest for excitation at the single particle plasmon frequency, occurring at the center of the first Brillouin zone. Corrections to this solution by considering higher order multipoles - albeit still in the quasi-static approximation - have also been obtained [Park and Stroud, 2004].

Solutions for the dispersion relations using the full set of Maxwell's equations, thus overcoming the quasi-static approximation, have revealed a significant change in the dispersion relation for the transverse mode near the light line (Fig. 7.22), due to phase-matching between the transverse dipolar mode to photons propagating along the waveguide at the same frequency [Weber and Ford, 2004, Citrin, 2005b, Citrin, 2004]. For longidutinal modes, this coupling cannot take place, and the obtained curves are similar to the quasistatic result. Examples of the electric field distribution of the guided modes are shown in Fig. 7.23, which depicts results from finite-difference time-domain simulations of pulse propagation through a chain of 50 nm gold spheres separated by a center-to-center distance of 75 nm in air. These simulations have also confirmed the negative phase velocity of transverse modes [Maier et al., 2003a].

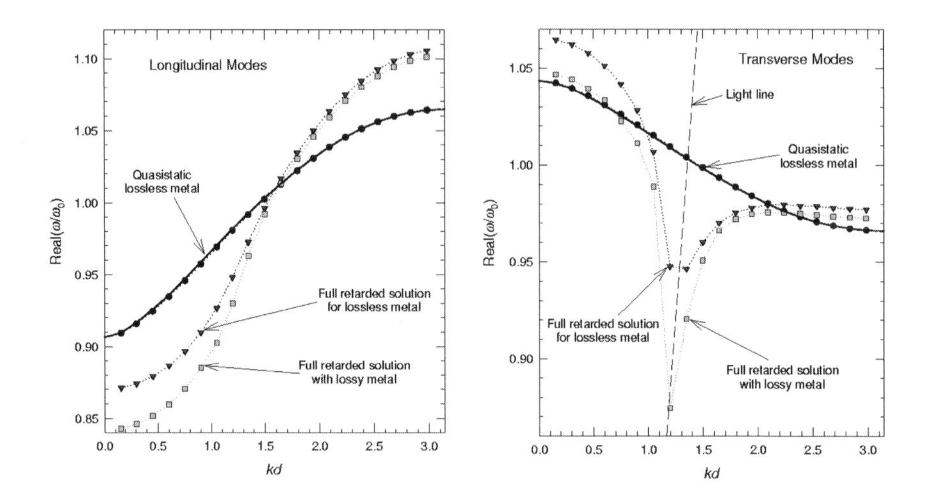

Figure 7.22. Dispersion of longitudinal (left panel) and transverse (right panel) modes sustained by an infinite chain of spherical particles in the quasi-static approximation (solid lines, [Brongersma et al., 2000]), for a finite 20-sphere chain in the quasi-static approximation (full circles), and for the fully retarded solution with a lossy metal (squares) and for a lossless metal (triangles). Differences between the models are pronounced for transverse polarization. Reprinted with permission from [Weber and Ford, 2004]. Copyright 2004 by the American Physical Society.

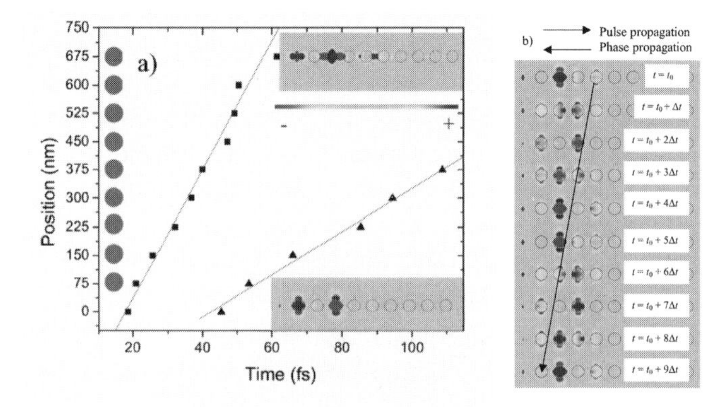

Figure 7.23. Finite-difference time-domain simulation of a pulse propagating through a chain of 50 nm gold spheres with a 75 nm center-to-center distance. (a) Position of the peak of a pulse centered around the single particle resonance frequency with time as it propagates through the chain for longitudinal (squares) and transverse (triangles) polarization. The insets show snapshots of the electric field distribution. (b) Snapshots of the electric field distribution of a transverse mode traveling with negative phase velocity. The arrow denotes the movement of a particular phase front. Reprinted with permission from [Maier et al., 2003a]. Copyright 2003 by the American Physical Society.

The excitation of traveling waves at the point of highest group velocity requires a local excitation scheme, since far-field excitation only excites modes around the $\mathbf{k} = 0$ point in the dispersion diagram. By analysing the shift of the plasmon resonance compared to that of a single particle (or an array of sufficiently separated particles), due to interparticle coupling upon in-phase excitation (as presented in chapter 5), the strength of the coupling can be judged. Fig. 7.24 shows as an example a waveguide consisting of silver rods of aspect ratio $90 \times 30 \times 30$ nm^3 separated by a gap of 50 nm, and far-field extinction spectra of the chain as well as of well-separated particles. A significant blue-shift shift due to particle coupling is apparent for the chain.

In order to locally excite a traveling wave on this structure, the tip of a near-field optical microscope was used as a local illumination source, and the energy transport along the particle array detected via fluorescent polymer beads (Fig. 7.25a) [Maier et al., 2003b]. In this study, the tip of the near-field microscope was scanned over an ensemble of waveguides (panel b), and the recorded fluorescent spots in the obtained near-field images compared between beads situated at a distance from the waveguides (panel c) and those deposited on top of them (panel d). The latter showed an elongation of the spot profile along the direction of the waveguide due to excitation at a distance via the particle waveguide: energy is transfered from the tip to the waveguide, and channeled to the fluorescent particle (see scheme in panel a). Representative cross cuts

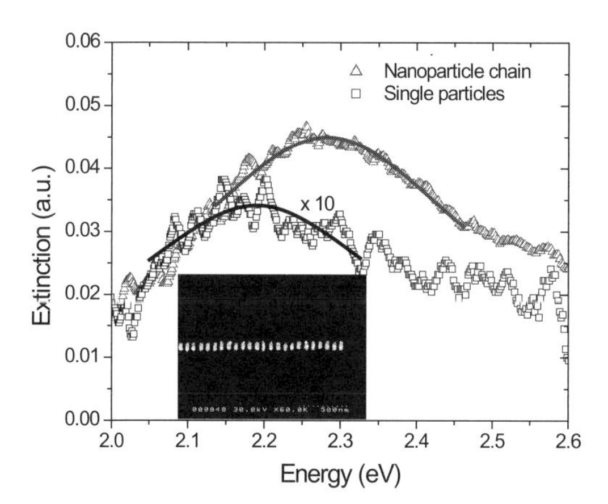

Figure 7.24. Experimentally observed plasmon resonance for single silver rods and a chain of closely-spaced rods under transverse illumination (along the long axis of the rods). The blue-shift between the two spectra is due to near-field interactions between particles in the chain. Reprinted by permission of Macmillan Publishers Ltd: Nature Materials [Maier et al., 2003b], copyright 2003.

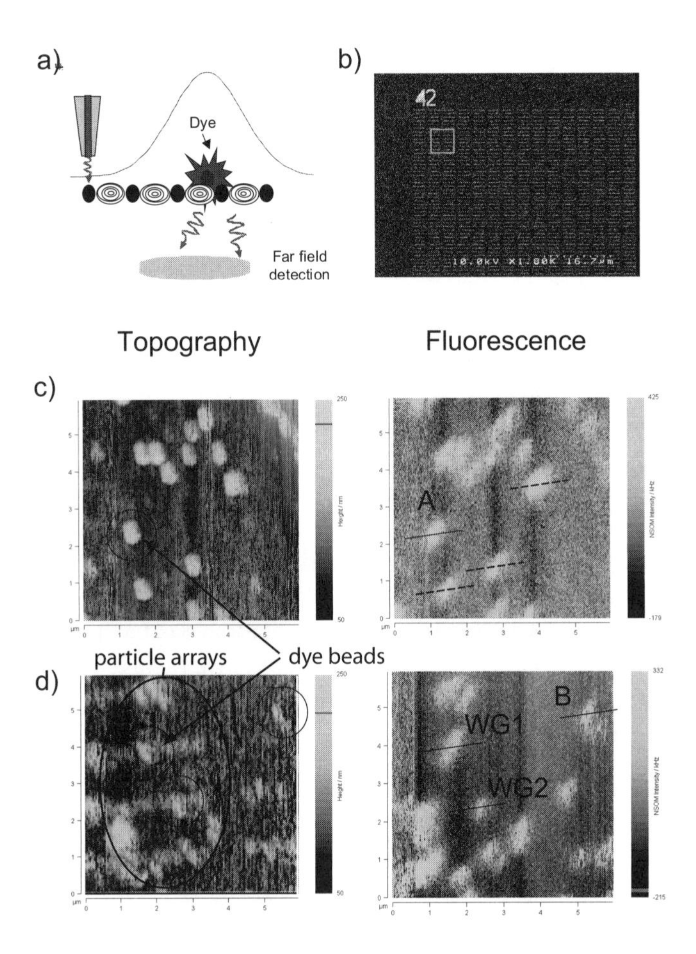

Figure 7.25. Local excitation and detection of energy transport in metal nanoparticle plasmon waveguides. Schematic of the experiment (a), SEM images of plasmon waveguides (b), and images of the topography and fluorescence (c, d). The images presented in (c) show fluorescent spheres deposited in a region without waveguides, while (d) shows spheres deposited on top of the ends of four nanoparticle chains. The circles and lines mark the fluorescent spots analyzed in Fig. 7.26. Reprinted by permission of Macmillan Publishers Ltd: Nature Materials [Maier et al., 2003b], copyright 2003.

through the fluorescent spots are shown in Fig. 7.26, suggesting energy transport along the particle chain over a distance of 500 nm. A numerical analysis has confirmed the major aspects of this coupling scheme [Girard and Quidant, 2004].

Due to the resonant excitation at the particle plasmon resonance frequency, the fields are highly confined to the waveguide structure, akin to the nanowires presented in the preceding section. This implies high losses, with propagation lengths on the order of 1 μm or below, depending on the wavelength of opera-

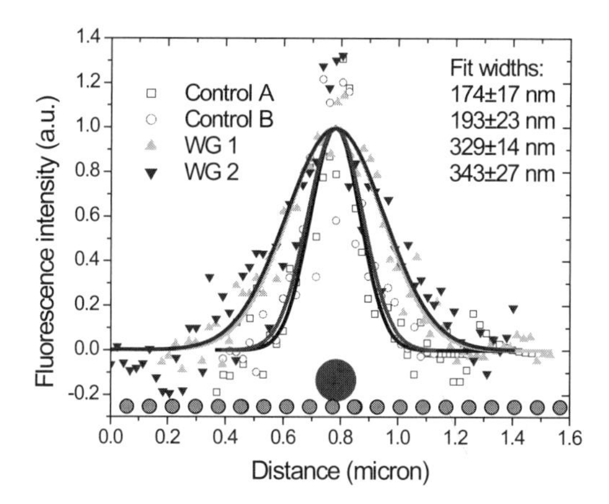

Figure 7.26. Intensity of the fluorescence signal along the cross-cuts indicated in Fig. 7.25c,d for control fluorescent spheres located away from waveguides (squares) and spheres located on top of particle waveguides (triangles). The increase in width of the fluorescence peaks for the latter is indicative of excitation at a distance via the particle waveguides (see sketch in Fig. 7.25a). Reprinted by permission of Macmillan Publishers Ltd: Nature Materials [Maier et al., 2003b], copyright 2003.

tion and the dielectric constant of the surrounding host material. Applications such as condensers for channeling energy have been demonstrated [Nomura et al., 2005], and the possibility to use short structures of self-similar spheres as nanolenses for near-field focusing akin to the conical tapers presented in the previous section has been suggested [Li et al., 2003].

Significantly longer propagation lengths can be achieved by using non-resonant particle excitation at lower frequencies. However, while the absorptive losses are lowered, now radiative losses begin to overwhelm the guiding, and a different approach than one-dimensional chains is needed to keep the energy confined to the waveguide. A promising approach to achieve this was demonstrated in the form of a nanoparticle plasmon waveguide operating in the telecommunications window at $\lambda_0 = 1.5\ \mu$m [Maier et al., 2004, Maier et al., 2005]. The new design exhibited a confinement superior to the long-ranging stripe waveguides discussed in section 7.3, while still exhibiting propagation lengths of the order of 100 μm. The waveguide is based on a two-dimensional lattice of metal nanoparticles on a thin, undercut silicon membrane (Fig. 7.27d). Vertical confinement is achieved by a hybrid plasmon/membrane-waveguide mode, while transverse confinement can be achieved by using a lateral grading of nanoparticle size, thus in a sense creating a higher effective refractive index in the waveguide center. This way, the mode is confined to the higher-index region, leading to wavelength-scale transverse

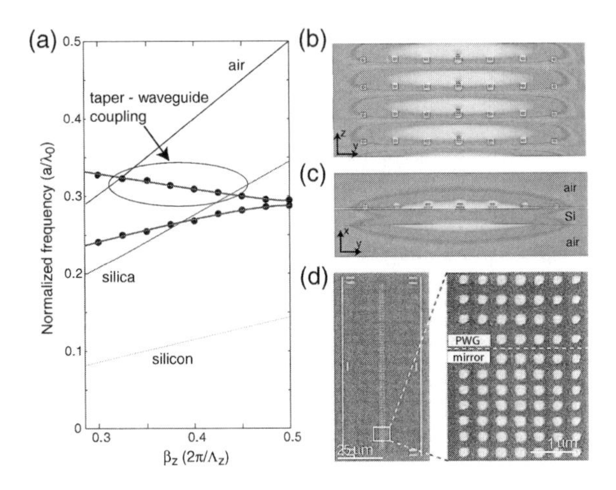

Figure 7.27. Dispersion relation (a) and mode profiles in top (b) and side (c) view of a metal nanoparticle plasmon waveguide on a thin Si membrane operating in the near-infrared. (d) SEM picture of a fabricated device. Reprinted with permission from [Maier et al., 2004]. Copyright 2004, American Institute of Physics.

confinement and sub-wavelength scale vertical confinement (Fig. 7.27b,c). We point out that this concept for engineering the electromagnetic response via a particle lattice is akin to that of designer plasmons presented in chapter 6.

Due to the periodicity in the propagation direction, the plasmon mode is zone-folded back into the first Brillouin zone (Fig. 7.27a). This suggests a convenient scheme for excitation using fiber tapers placed on top of the waveguide (see Fig. 3.14): contra-directional phase-matched evanescent coupling between the fiber taper and the plasmon mode can take place.

The fiber taper is also a convenient means to investigate both the spatial and the dispersive properties of the nanoparticle waveguide. For a spatial mapping of the guided modes, the fiber simply has to be moved over the waveguide in the transverse direction, and the wavelength-dependent power transferred past the coupling region monitored. As an example, Fig. 7.28a shows the power transmitted past the coupling region vs. wavelength and transverse location of the taper over the waveguide. Both the fundamental and the first higher-order mode of the plasmon waveguide manifest themselves via power drops at 1590 nm and 1570 nm (Fig. 7.28b,c), depending on whether the taper is located over the waveguide center or at its edges. The spatial resolution is of course limited by the diameter of the taper, which in this case was about 1.5 μm.

Translation of the taper in the direction of the waveguide moves the point of phase-matching via a change in taper diameter. This can be used to map out the dispersion relation, and confirm the contra-directional nature of the coupling (Fig. 7.29a): As the diameter of the taper is increased (and thus its dispersion

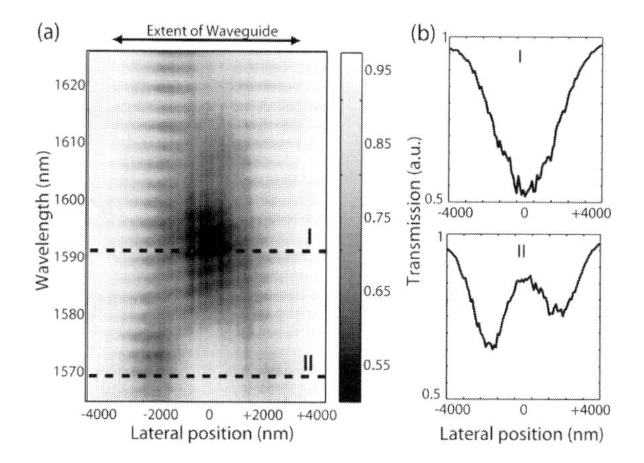

Figure 7.28. (a) Power transferred past the coupling region vs. wavelength and transverse taper position. Both the fundamental and the first-order mode are shown in data cuts (b). Reprinted with permission from [Maier et al., 2005]. Copyright 2005, American Institute of Physics.

curve moves closer to the silica light line), the point of phase-matching shows a red-shift. A look at the dispersion diagram of Fig. 7.27a confirms that this is only the case for coupling to the zone-folded upper band. The maximum power transfer efficiency demonstrated experimentally using this geometry is about 75% (Fig. 7.29b).

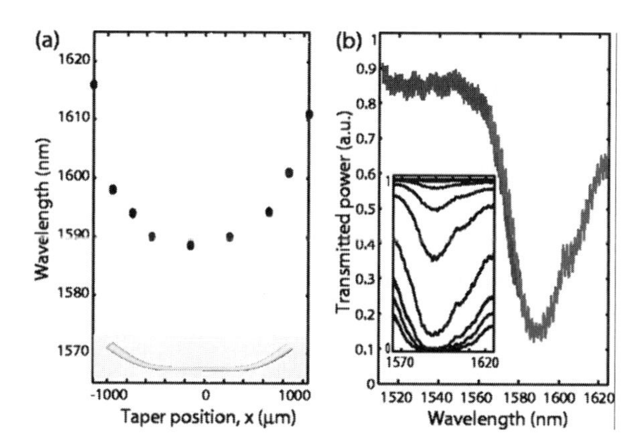

Figure 7.29. (a) Spectral position of the phase matching point vs. taper position as the taper is moved along the waveguide axis, demonstrating the contra-directional nature of the coupling. (b) Power transmitted past the coupling region for the condition of optimum coupling, demonstrating transfer efficiencies of about 75%. The inset shows the evolution of the coupling profile as the gap between the taper and the waveguide is descreased. Reprinted with permission from [Maier et al., 2005]. Copyright 2005, American Institute of Physics.

These low-loss metal nanoparticle waveguides could be employed in applications for coupling radiation transmitted through optical fibers into two-dimensional SPP modes with high efficiency. After the coupling region, guiding to desired structures on a chip for sensing is possible, perhaps after interfacing with higher-confinement waveguides for field focusing.

7.7 Overcoming Losses Using Gain Media

We have already discussed in chapter 5 the possibility of overcoming the inherent attenuation losses (due to Ohmic heating) in metallic structures by embedding them into media with optical gain. For particles, optical gain results in an increase of the magnitude of the polarization (5.7) and a concomitant decrease in the linewidth of the resonant mode, limited by gain saturation. Applied to waveguides, an analytical study of particle chains (akin to the nanoparticle plasmon waveguides discussed in the preceding section) embedded in a gain medium suggests that the accompanying increase in interparticle coupling strength can lead to greatly enhanced propagation distances, particularly for confined transverse modes close to the light line [Citrin, 2005a].

In the wider context of waveguiding using propagating SPPs at flat interfaces, one can therefore expect that the presence of gain media will result in an increase of the propagation length L. More surprisingly, it can also easily be shown that the localization of the fields to the interface will be increased [Avrutsky, 2004], contrary to the trade-off between confinement and loss present in the absence of gain. To demonstrate this, one can define the effective index of the SPP at an interface between a metal and a dielectric via the dispersion relation (2.14) as

$$n_{\text{eff}} = \sqrt{\frac{\varepsilon\varepsilon_d}{\varepsilon + \varepsilon_d}}, \tag{7.3}$$

where ε_d is the permittivity of the insulating layer. As in the discussion of localized plasmons, we see that in the resonant limit of surface plasmons, defined by $\text{Re}\,[\varepsilon] = -\varepsilon_d$, the effective index and thus the amount of localization is limited by the non-vanishing imaginary part of ε due to attenuation. However, in analogue to the discussion in chapter 5, the presence of gain can lead to a complete vanishing of the denominator of (7.3), and thus a large effective index (limited only by gain saturation).

While the effect of this increase in n_{eff} on SPP propagation in waveguides has not been analyzed in detail up to this point, various analytical and numerical studies have focused on the increase in propagation length offered, both for metal stripe [Nezhad et al., 2004] and gap waveguides [Maier, 2006a]. For both geometries with excitation at near-infrared frequencies, the gain coefficients required for lossless propagation are at the boundary of what is currently achievable using quantum-well or quantum-dot media. Taking a simple

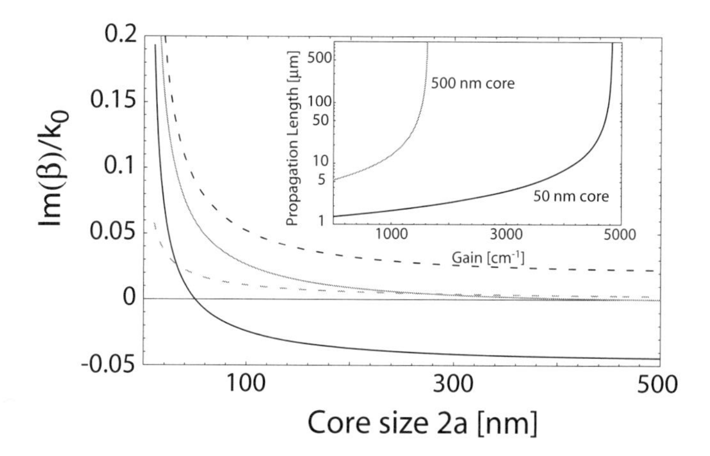

Figure 7.30. Evolution of the imaginary part of the propagation constant $\text{Im}\left[\beta\right]$ of a gold/dielectric/gold waveguide with decreasing core size for cores consisting of air (broken gray line), a semiconductor material ($n = 3.4$) with zero gain (broken black line), and gain coefficients $\gamma = 1625\ \text{cm}^{-1}$ (gray line) and $\gamma = 4830\ \text{cm}^{-1}$ (black line), respectively. The insets show the energy propagation length of the mode. As the critical gain for which $\text{Im}\left[\beta\right] = 0$ is approached, $L \rightarrow \infty$ (inset). Reprinted from publication [Maier, 2006a], copyright 2006, with permission from Elsevier.

one-dimensional gold-semiconductor-gold gap waveguide as an example, lossless propagation at $\lambda_0 = 1500$ nm for a core size of only 50 nm is expected for a gain coefficient $\gamma = 4830\ \text{cm}^{-1}$ in a core layer with $n = 3.4$. This is demonstrated in Fig. 7.30, which shows the evolution of the imaginary part of the propagation constant $\text{Im}\left[\beta\right]$ with decreasing core size for waveguides with cores consisting of air (broken gray line) or a semiconductor material ($n = 3.4$) with zero gain (broken black line), or gain coefficients $\gamma = 1625\ \text{cm}^{-1}$ (gray line) or $\gamma = 4830\ \text{cm}^{-1}$ (black line), respectively. Note that $\text{Im}\left[\beta\right] < 0$ implies an exponential increase of the energy of the guided wave. As expected, the propagation distance increases with the amount of gain present, shown in the inset.

After these promising theoretical studies, it remains to be seen if the large gain coefficients necessary for low-loss or even lossless propagation are indeed achievable in close vicinity of metallic guiding structures.

Chapter 8

TRANSMISSION OF RADIATION THROUGH APERTURES AND FILMS

Up to this point, our discussion of surface plasmon polaritons has focused on their excitation and guiding *along* a planar interface. In the previous chapter, we have seen how control over the propagation of these two-dimensional waves for waveguiding applications can be achieved by surface patterning. Here, we move in the perpendicular direction and take a look at the transmission of electromagnetic energy *through* thin metallic films, aided by near-field effects. If the film is patterned with a regular array of holes, or surface corrugations surrounding a single hole, phenomena such as enhanced transmission and directional beaming can occur, which have triggered an enormous amount of interest ever since their first description in 1998.

To lay the foundations for the discussion of these effects, we begin by reviewing the basic physics of the transmission of light through a sub-wavelength circular hole in a thin conductive screen. Subsequent chapters treat the transmission enhancement encountered in hole arrays and the directional control over the transmitted beam via surface corrugations at the exit side of the interface. The role of SPPs and localized plasmons in the transmission of light through a single hole surrounded by regular corrugations is also addressed. The chapter closes with a look at first applications of these effects and a discussion of light transmission through unperforated films mediated by coupled SPPs.

8.1 Theory of Diffraction by Sub-Wavelength Apertures

The physics of the transmission of light through a single hole in an opaque screen, also called an aperture, has been a topic of intense research for well more than a hundred years. Due to the wave nature of light, its transmission through an aperture is accompanied by *diffraction*. Therefore, this process, which even in the simplest of geometries is very complex, can be described using various approximations developed in classical diffraction theory. A review

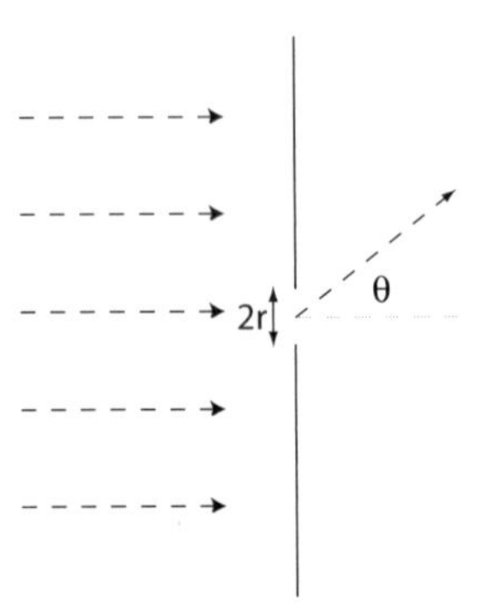

Figure 8.1. Transmission of light through a circular aperture or radius r in an infinitely thin opaque screen.

of different aspects of this theory can be found in basic textbooks on electrodynamics such as [Jackson, 1999], and (from the point of view of the transmission problem presented in this chapter) in the review article by Bouwkamp [Bouwkamp, 1954]. A geometry that has received particular attention in these treatments, due to its relative easy tractability, is that of a circular aperture of radius r in an infinitely thin, perfectly conducting screen (Fig. 8.1).

For an aperture with a radius r significantly larger than the wavelength of the impinging radiation ($r \gg \lambda_0$), this problem can be treated quite successfully using the Huygens-Fresnel principle and its mathematical formulation, the *scalar* diffraction theory by Kirchhoff [Jackson, 1999]. Since this theory is based on the scalar wave equation, it does not take into account effects due to the polarization of light. For normally-incident plane-wave light, it can be shown that the transmitted intensity per unit solid angle in the far field (known as the limit of *Fraunhofer diffraction*) is given by

$$I(\theta) \cong I_0 \frac{k^2 r^2}{4\pi} \left| \frac{2J_1(kr\sin\theta)}{kr\sin\theta} \right|^2, \tag{8.1}$$

where I_0 is the total intensity impinging on the aperture area πr^2, $k = 2\pi/\lambda_0$ the wavenumber, θ the angle between the aperture normal and the direction of the re-emitted radiation, and $J_1(kr\sin\theta)$ the Bessel function of the first kind. The functional form described by (8.1) is that of the well-known Airy pattern of a central bright spot surrounded by concentric rings of decreasing intensity, caused by angle-dependent destructive and constructive interference of rays originating from inside the aperture. The ratio of the total transmitted intensity to I_0, given by

$$T = \frac{\int I(\theta)\, d\Omega}{I_0}, \tag{8.2}$$

is called the *transmission coefficient*. For apertures with $r \gg \lambda_0$, in which case the treatment outlined here is valid, $T \approx 1$. In this regime, more exact calculations of the diffraction problem give semi-quantitatively essentially the same result as (8.1).

Since we are interested in the influence of surface waves such as SPPs on the transmission process, the regime of sub-wavelength apertures $r \ll \lambda_0$ is much more interesting, because near-field effects are expected to dominate the response (due to the absence of propagating modes in apertures in films of *finite* thickness). However, even an approximate analysis of an infinitely thin perfectly conducting screen requires an approach using the full vectorial description via Maxwell's equations. The basic assumption of Kirchhoff's method is that the electromagnetic field in the aperture is the same as if the opaque screen were not present, which does not fulfill the boundary condition of zero tangential electric field on the screen. For large holes, this basic failure is less severe, since the diffracted fields are relatively small compared to the directly-transmitted field. For sub-wavelength apertures on the other hand, this approximation is inadequate even as a first-order treatment of the problem.

Assuming that the incident light intensity I_0 is constant over the area of the aperture, Bethe and Bouwkamp arrived at an exact analytical solution for light transmission through a sub-wavelength circular hole in a perfectly conducting, infinitely thin screen [Bethe, 1944, Bouwkamp, 1950a, Bouwkamp, 1950b]. For normal incidence, the aperture can be described as a small *magnetic dipole* located in the plane of the hole. The transmission coefficient for an incident plane wave is then given by

$$T = \frac{64}{27\pi^2}(kr)^4 \propto \left(\frac{r}{\lambda_0}\right)^4. \tag{8.3}$$

The scaling with $(r/\lambda_0)^4$ implies very weak total transmission (smaller by an amount of the order of $(r/\lambda_0)^2$ compared to Kirchhoff theory) for a sub-wavelength aperture, as can intuitively be expected. Also, the scaling $T \propto \lambda_0^{-4}$ is in agreement with Rayleigh's theory of the scattering by small objects. We note that (8.3) is valid for normally-incident radiation both in TE and TM polarization. For radiation impinging on the aperture at an angle, an additional *electric dipole* in the normal direction is needed to describe the transmission process. In this case, more radiation is transmitted for TM than for TE polarization [Bethe, 1944].

The Bethe-Bouwkamp description of transmission through a circular aperture in a screen relies on two major approximations. The thickness of the conducting screen is assumed to be infinitely thin, yet the screen is still perfectly opaque due to the infinite conductivity. Relaxing the first assumption and thus treating screens of finite thickness h requires numerical simulations for solving of the problem. Two regimes have to be considered, depending on whether the waveguide defined by the sub-wavelength aperture allows a propagating mode to exist or not. The Bethe-Bouwkamp model is only applicable to apertures which allow only decaying modes. For a circular (square) hole of diameter d in a perfect screen, this condition is fulfilled in the regime where $d \lesssim 0.3\lambda_0$ ($d \leq \lambda_0/2$), which can be calculated via a boundary analysis at the rim of the aperture waveguide. The transmission coefficient T then decreases exponentially with h [Roberts, 1987]. This is of course the behavior characteristic of a tunneling process. For sub-wavelength apertures allowing propagating modes, the theory outlined here is not applicable and T is much higher due to the waveguide behavior of the aperture. Prominent examples of such *waveguide apertures* are circular holes with diameters above the cut-off [de Abajo, 2002], the well-known one-dimensional slit (which has a TEM mode without cut-off), annular-shaped apertures [Baida and van Labeke, 2002], and apertures in the form of a C-shape [Shi et al., 2003].

Apart from the finite screen thickness, when discussing the transmission properties of real apertures the finite conductivity of the metal screen should be taken into account. For optically thin films, the screen is thus not perfectly opaque, and comparisons with the Bethe-Bouwkamp theory are not justified. On the other hand, an optically thick film of a real metal satisfies the condition of opacity if h is on the order of several skin depths, thus preventing radiation tunnelling through the screen. For apertures fulfilling this condition, it has been shown that localized surface plasmons significantly influence the transmission process [Degiron et al., 2004]. This will be discussed in more detail in a later section, after a description of the role of SPPs excited via phase-matching on the input side of the screen in the tunneling process.

8.2 Extraordinary Transmission Through Sub-Wavelength Apertures

The transmission of light through a sub-wavelength aperture of a geometry such as a circle or a square that does not allow a propagating mode can be dramatically enhanced by structuring the screen with a regular, periodic lattice. This way, SPPs can be excited due to grating coupling, leading to an enhanced light field on top of the aperture. After tunneling through the aperture, the energy in the SPP field is scattered into the far field on the other side.

The phase-matching condition imposed by the grating leads to a well-defined structuring of the transmission spectrum $T(\lambda_0)$ of the system, with peaks at the wavelengths where excitation of SPPs takes place. At these wavelengths, $T > 1$ is possible - more light can tunnel through the aperture than incident on its area, since light impinging on the metal screen is channeled through the aperture via SPPs. This *extraordinary transmission* property was first demonstrated by Ebbesen and co-workers for a square array of circular apertures in a thin silver screen [Ebbesen et al., 1998].

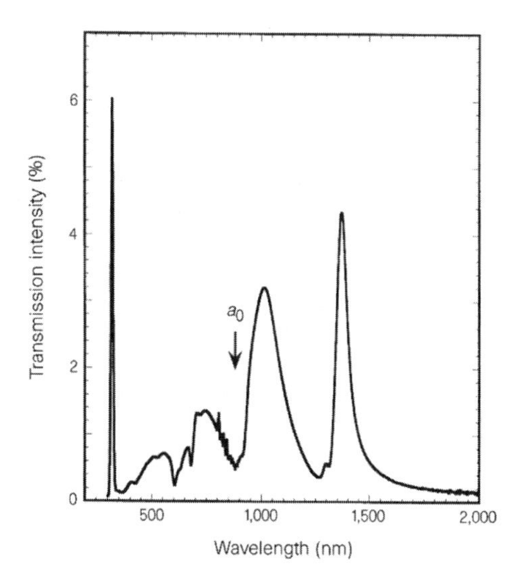

Figure 8.2. Normal-incidence transmission spectrum for a silver screen perforated with a square array of holes of diameter $d = 150$ nm and lattice constant $a_0 = 900$ nm. The thickness of the screen is 200 nm. Reprinted by permission from Macmillan Publishers Ltd: Nature [Ebbesen et al., 1998], copyright 1998.

As a typical example, Fig. 8.2 shows the transmission spectrum for normally-incident light on a silver screen of thickness $t = 200$ nm perforated with an array of circular holes of diameter $d = 150$ nm arranged on a square lattice with period $a_0 = 900$ nm. Apart from a sharp peak in the ultraviolet region only observable for very thin films, the spectrum shows a number of distinct, relatively broad peaks, two of which occur at wavelengths above the grating constant a_0. The origin of these peaks cannot be explained by a simple diffraction analysis without assuming the contribution of surface modes, and the fact that $T > 1$ suggests that the transmission is mediated via SPPs excited via grating-coupling at the periodic aperture lattice: This way, also light impinging on opaque regions between the apertures can be channeled to the other side via propagating SPPs. We note however that experimentally the exact determination of the transmission enhancement is difficult, due to the problem of normalization: the transmission calculated using the Bethe formula (8.3) requires a highly accurate determination of the aperture dimensions, due to the strong dependence of $T \propto r^4$ on the aperture radius. We note that in the initial studies, the normally-incident light was not polarized, and that in fact due to the square symmetry of the aperture arrays identical transmission spectra occur for TM and TE polarization [Barnes et al., 2004].

A study of the dependence of the peak positions on incidence angle of the radiation allows the mapping of the dispersion relation of the waves involved in

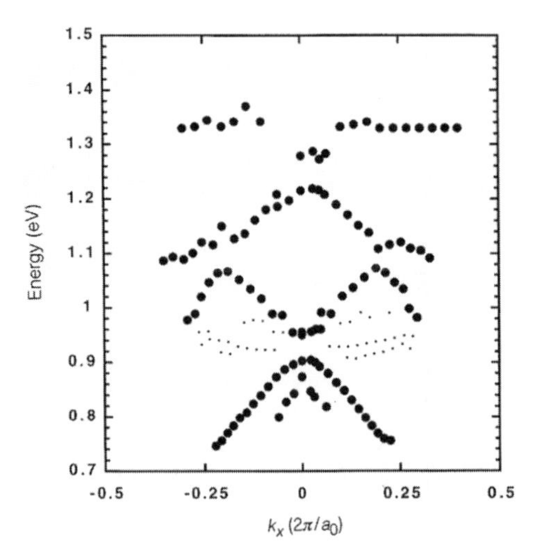

Figure 8.3. Dispersion relation of grating-coupled SPPs along the [10] direction of the aperture array extracted from spectra such as Fig. 8.2 for different incidence angles (solid dots). Reprinted by permission from Macmillan Publishers Ltd: Nature [Ebbesen et al., 1998], copyright 1998.

the transmission process. An example is shown in Fig. 8.3. The typical form of the SPP dispersion relation (2.14), displaced by the grating vector $G = 2\pi/a_0$, can be clearly discerned. The crossing of the dispersion curves with the $k_x = 0$-axis defines the points of phase-matching for normal-incidence of the exciting light beam, and thus the position of the transmission maxima in Fig. 8.2.

The observed structure of $T(\lambda)$ can therefore be explained by assuming that grating coupling to SPPs takes place, with the phase matching condition

$$\beta = k_x \pm nG_x \pm mG_y = k_0 \sin\theta \pm (n+m)\frac{2\pi}{a_0}, \tag{8.4}$$

where β is the SPP propagation constant. For phase-matching via a square lattice, it can easily be shown by combining (8.4) and (2.14) that for normally-incident light the transmission maxima occur at wavelengths fulfilling the condition [Ghaemi et al., 1998]

$$\lambda_{\mathrm{SPP}}(n,m) = \frac{n_{\mathrm{SPP}} a_0}{\sqrt{n^2+m^2}}. \tag{8.5}$$

$n_{\mathrm{SPP}} = \beta c/\omega$ is the effective index of the SPP, which is for the single interface between a metal and a dielectric calculated using (2.14). This simplified description often serves as a good first approximation.

Since phase-matching of the incident radiation to SPPs is crucial for transmission enhancement via SPP tunneling, the same process should occur for a single hole surrounded by a regular array of opaque surface corrugations. This was indeed confirmed in a follow-up study with only one aperture, where dimples instead of holes in the screen served as the grating for coupling [Grupp et al., 1999]. Apart from a two-dimensional square lattice of apertures or dimples, concentric circles surrounding the aperture can also be used to achieve phase-matching of the incident light beam with SPPs. Fig. 8.4 shows transmission spectra for such a *bull's eye* structure with sets of concentric rings of different groove height h (A), and also for a two-dimensional dimple array (B). Transmission enhancement compared to the values calculated using (8.3) is present in both cases, and for the bull's eye structure additionally $T > 1$ occurs at the wavelengths of phase-matching. It is apparent from Fig. 8.4a that the height h of the undulations responsible for coupling determines the efficiency of SPP coupling, and therefore the magnitude of the transmission enhancement.

We now want to qualitatively describe the physics of the transmission process in more detail. Similar to a single aperture in an unpatterned screen, transmission through an aperture in a regularly patterned surface occurs via tunneling, leading to an approximately exponential dependence of the transmitted intensity on the thickness t of the metal screen. However, if t is of the order of the skin depth, coupling between SPPs at the front and back interface takes place

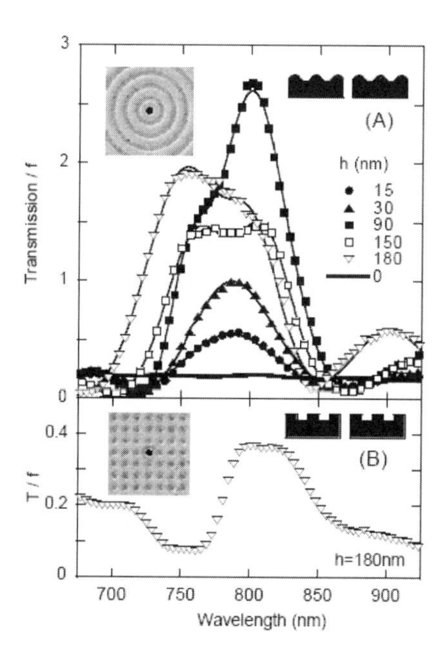

Figure 8.4. Transmission through a single circular aperture (d = 440 nm) surrounded by concentric rings with sinusoidal cross section (A) or a square array of dimples (B) of height h milled into a 430 nm thick Ag/NiAg screen. Reprinted with permission from [Thio et al., 2001]. Copyright 2001, Optical Society of America.

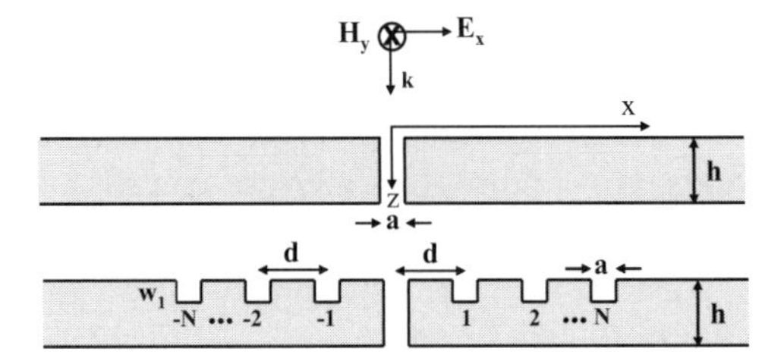

Figure 8.5. Schematic of a single slit aperture cut into a perfectly conducting screen, with and without a surrounding groove array on the input side. Courtesy of Francisco García-Vidal, Universidad Autónoma de Madrid. Figure similar to that in [García-Vidal et al., 2003a], Copyright 2003 by the American Physical Society.

if the adjacent dielectric media are equal, enabling phase-matching. Degiron and co-workers have shown that this leads to a saturation of the transmission coefficient for small screen thickness [Degiron et al., 2002]. A great number of studies have since then either experimentally or numerically studied the influence of geometrical parameters such as metal film thickness [Shou et al., 2005], hole size [van der Molen et al., 2004] or symmetry of the hole arrays [Wang et al., 2005] on the transmission spectra. Crucially, a comprehensive, polarization-resolved study of the angular dependence of the transmission, reflection and absorption of light by a metal film perforated with an array of sub-wavelength holes has confirmed the role of SPPs, excited via diffraction of the impinging light beam in the transmission process [Ghaemi et al., 1998, Barnes et al., 2004].

The complexity of the transmission process significantly increases for apertures allowing a propagating mode, such as an essentially one-dimensional slit structure, where the fundamental TEM mode does not exhibit a cut-off width. In this case, the transmission can be modulated via resonances of the fundamental slit waveguide mode, controlled by the thickness of the metal film. Transmission resonances have indeed been observed for arrays of parallel, sub-wavelength slits [Porto et al., 1999]. In analogy with the discussion of extraordinary transmission through apertures via tunneling, periodic surface corrugations around a single slit, shown in Fig. 8.5, significantly increase the transmission and allow $T(\lambda) > 1$ due to the excitation of SPPs.

The fact that even perfect metals can sustain surface waves in the form of designer plasmons on patterned interfaces as described in chapter 6 leads to enhanced transmission phenomena also in this limit. Using a modal expansion technique similar to that presented in chapter 6 for describing designer SPPs at low frequencies, García-Vidal and co-workers have shown that the transmission spectrum $T(\lambda)$ for a slit aperture surrounded by parallel grooves is additionally influenced by the existence of coupled cavity modes in the grooves, the frequencies of which are defined by their depth h. Also, in-phase re-emission from the array, controlled via the period d, takes place [García-Vidal et al.,

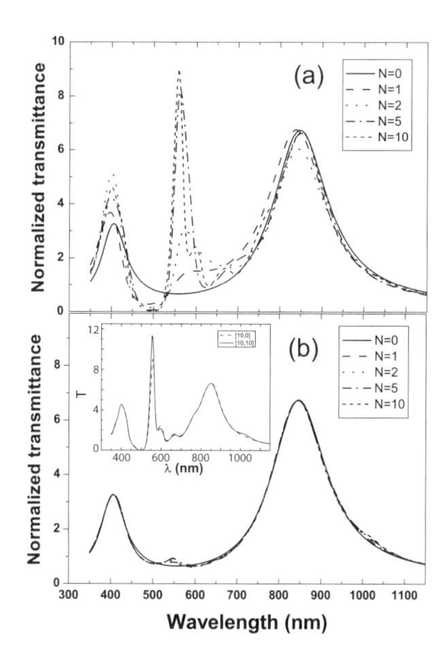

Figure 8.6. Normalized transmittance $T(\lambda)$ of the slit structure depicted in Fig. 8.5 for a = 40 nm, d = 500 nm, w = 350 nm and groove depth h = 100 nm. The number of grooves patterned either on the input side (a) or the exit side (b) is $2N$. Transmission enhancement is only present for patterning of the input side (a), and only a small number of grooves are necessary to establish a significant enhancement, while patterning of the exit surface does not lead to (b) nor significantly influence (inset in b) the magnitude of $T(\lambda)$. Courtesy of Francisco García-Vidal, Universidad Autónoma de Madrid. Figure similar to that in [García-Vidal et al., 2003a], Copyright 2003 by the American Physical Society.

2003a, Marquier et al., 2005]. Fig. 8.6 shows theoretical calculations of the dependence of $T(\lambda)$ on the number of grooves based on this model (a), further demonstrating that only surface patterning on the entrance side has significant influence on the maxima in $T(\lambda)$ (b).

For the geometrical parameters chosen in this calculations, the two transmission maxima around 400 nm and 850 nm not influenced by the patterning correspond to slit waveguide resonances, while the strong and sharp maximum at λ = 560 nm is due to the establishment of groove cavity modes and in-phase groove reemission with increasing number of grooves, mediated by designer surface plasmons. The main results of this study have been confirmed independently by using a different approach based on scattering theory from quantum mechanics [Borisov et al., 2005]. Furthermore, control of the phase of the re-emitted radiation allows selective suppression of transmission, as has been confirmed with suitable phase-gratings in the THz regime [Cao et al., 2005]. A recent study has further shown that even a one-dimensional array of sub-wavelength apertures exhibits many of the features present in the two-dimensional patterning studies [Bravo-Abad et al., 2004a]. We note that extraordinary transmission via excitation of SPPs has not only been observed for visible light using metallic screens, but also for highly doped semiconductors and polymer films at THz frequencies [Matsui et al., 2006].

While the patterning of the input surface of the aperture screen determines the spectral dependence $T(\lambda)$ of the transmission process, structuring of the exit surface allows control of the re-emission of the transferred radiation, which will be discussed in the next section.

8.3 Directional Emission Via Exit Surface Patterning

We have seen above that the tunneling of light through a sub-wavelength aperture below cut-off can be significantly enhanced by patterning the input side of the screen to allow phase-matching to SPPs. In a similar fashion, the emission on the exit side of the screen can be controlled via surface patterning as well. While not increasing $T(\lambda)$ (see Fig. 8.6b), imposing a regular grating

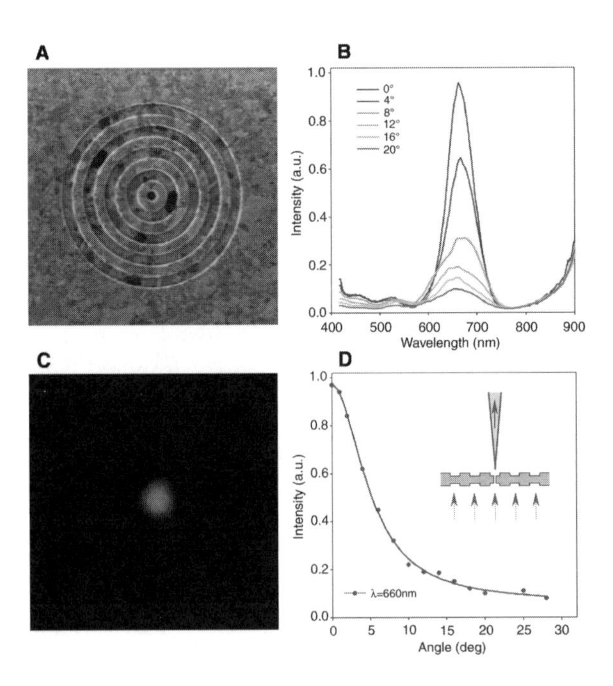

Figure 8.7. (a) Focused ion beam image of a bull's eye structure surrounding a circular subwavelength aperture in a 300 nm thick silver film. (b) Transmission spectra recorded at various collection angles, demonstrating the small divergence of the emerging beam (groove periodicity 600 nm, groove depth 60 nm, aperture diameter 300 nm). (c) Optical image of the directional emission at the wavelength of peak transmission. (d) Angular intensity distribution of the emitted beam at the wavelength of maximum transmission. Reprinted with permission from [Lezec et al., 2002]. Copyright 2002, AAAS.

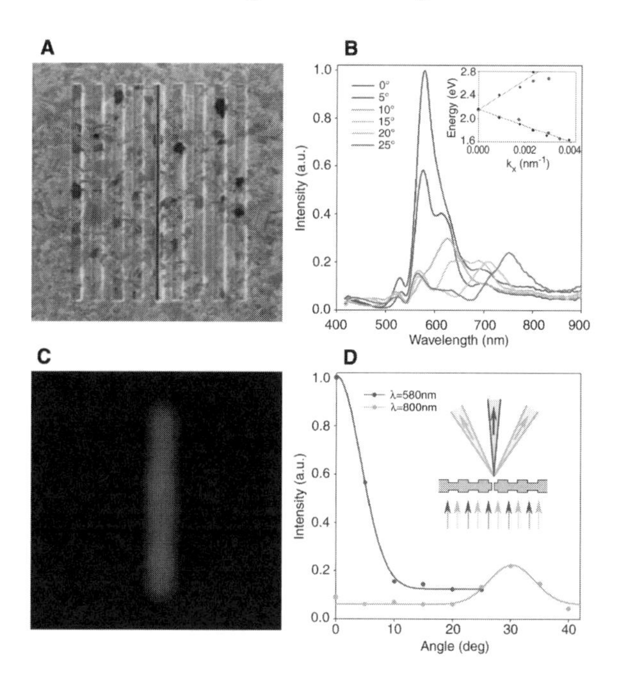

Figure 8.8. FIB image (A) and transmission spectrum for various collection angles (B) of a single sub-wavelength slit surrounded by parallel grooves cut into a 300 nm thick silver film (slit widh 40 nm, slit length 4400 nm, groove periodicity 500 nm, groove depth 60 nm). The inset in (B) shows the dispersion curve of the periodic structure (black dots) as well as the position of the spectral peaks (gray dots). (C) Optical image. (D) Angular intensity distribution of the emission at two selected wavelengths. Reprinted with permission from [Lezec et al., 2002]. Copyright 2002, AAAS.

structure on this side can lead to a highly directional emission with narrow beaming angle, first described by Lezec and co-workers [Lezec et al., 2002]. A patterning of both input and exit side of the screen can therefore lead to both enhanced transmission and directional emission.

Figures 8.7 and 8.8 show examples of this phenomenon for both a circular aperture surrounded by concentric grooves and a slit surrounded by a regular array of parallel grooves. The patterns is present on both sides of the film. While the position and amplitude of the transmission maxima $T(\lambda_0)$ are controlled by the phase-matching condition imposed by the pattern on the input side, the beam waist and direction of the emitted beam is governed by the exit side pattern. Highly directional emission with a angular divergence of approximately $\pm 3^\circ$ was observed. This phenomenon can be understood by assuming a SPP traveling from the exit side of the aperture along the screen towards the grooves and undergoing directional emission, defined by the groove period. Intriguingly, as a consequence light of different wavelengths can be emitted un-

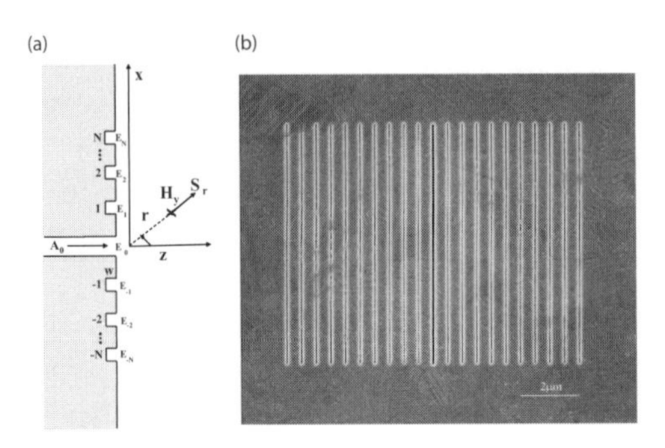

Figure 8.9. Schematic (a) and FIB image (b) of the exit surface of a screen with a single slit aperture surrounded by 10 parallel grooves on each side (slit width 40 nm, groove period 500 nm, and groove height 100 nm). Courtesy of Francisco García-Vidal, Universidad Autónoma de Madrid. Figure similar to that in [Martín-Moreno et al., 2003], Copyright 2003 by the American Physical Society.

der different angles (Fig. 8.8d), thus imposing a filtering property. In the study presented in these two figures, the groove periodicity was $d = 600$ nm or $d = 500$ nm, respectively, and the groove depth $h = 60$ nm. Fabrication was carried out using focused ion beam milling of a 300 nm thick free-standing silver film.

This intuitive picture of how the directional emission arises was corroborated by a theoretical analysis of the beaming profile for a slit aperture surrounded by a parallel array of grooves akin to that depicted in Fig. 8.8. The

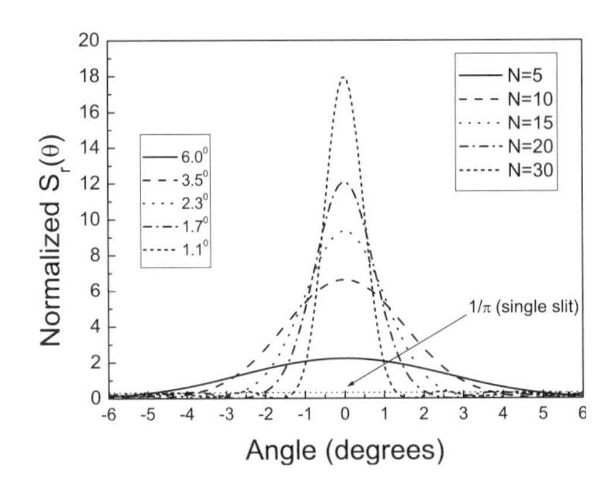

Figure 8.10. Theoretically predicted intensity profiles (angular intensity distribution) of the beam transmitted in the forward direction for the slit geometry of Fig. 8.9 for varying number $2N$ of grooves and geometrical parameters similar to those of Fig. 8.8. The legend also shows the angular divergence of the transmitted beam for each N. In the calculations, the groove depth has been adjusted for varying N in order to obtain similar total transmitted intensities. Courtesy of Francisco García-Vidal, Universidad Autónoma de Madrid. Figure similar to that in [Martín-Moreno et al., 2003], Copyright 2003 by the American Physical Society.

geometry of the system is defined in more detail in Fig. 8.9. Using a modal expansion of the fields in the slit and groove regions akin to the treatment described in chapter 6, Martín-Moreno and co-workers showed that beaming arises from tight-binding-like coupling between localized groove modes and the interference of their diffracted wave patterns [Martín-Moreno et al., 2003]. An example of the intensity profile $I(\theta)$ of the transmitted beam obtained using this model is shown in Fig. 8.10. A similar calculation for the exact parameters of the structure presented in Fig. 8.8 has demonstrated very good agreement between experiment and theory, both for the beam undergoing narrow transmission in the forward direction and the beam experiencing directional emission at an angle. Additionally, the theoretical treatment confirmed that only a small number of grooves $N \approx 10$ is needed for establishing the narrow beam profile.

The angular intensity distribution can thus be arranged almost at will by careful patterning of the exit surface of the screen, and it was even suggested that focusing at well-defined wavelengths could occur, with the screen effectively acting as a flat, wavelength-selective lens [García-Vidal et al., 2003b].

8.4 Localized Surface Plasmons and Light Transmission Through Single Apertures

As pointed out in the discussion of the limitations of the Bethe-Bouwkamp theory, even for an optically thick (and thus opaque) metal film the finite conductivity of the real metal should be taken into account to correctly analyze the transmission properties of a single aperture in a flat film. Penetration of the incident field inside the screen enables the excitation of *localized* surface plasmons on the rim of the aperture [Degiron et al., 2004], akin to the description in chapter 5 of localized modes in voids of a metallic film. One might expect that also propagating SPPs can be excited by viewing the aperture as a localized defect in the flat metal surface (see chapter 3). However, a detailed study of SPP excitation by a single aperture defect is still awaiting demonstration.

The excitation of localized surface plasmons at a single sub-wavelength aperture has two important consequences affecting the transmission $T(\lambda)$. Not surprisingly, due to the finite penetration of the fields into the rim of the aperture, its *effective diameter* is increased. This in turn leads to a substantial increase in the cut-off wavelength $\lambda_{\max}$ of the fundamental waveguide mode, compared to the physical diameter of the hole. Analytical and numerical studies have demonstrated an increase in $\lambda_{\max}$ of up to 41% [Gordon and Brolo, 2005], which has to be taken into account when studying apertures with a diameter just below the cut-off diameter for a perfectly-conducting screen, in light of the problem of correct normalization of the transmission coefficient mentioned earlier. Furthermore, theoretical studies of the transmission problem of a circular hole in a metal screen described by a free-electron dielectric

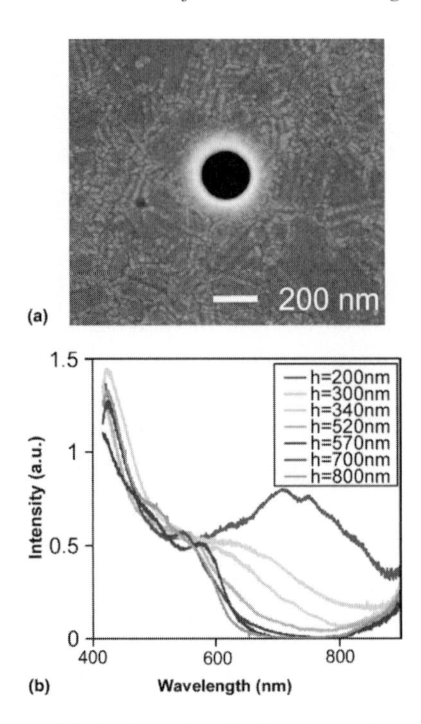

Figure 8.11. Transmission of light through a single sub-wavelength hole milled into a free-standing silver film (a). The transmission peak for small screen thickness h (b) is due to localized surface plasmons. Reprinted from [Degiron et al., 2004], copyright 2004, with permission from Elsevier.

function akin to (1.20) have suggested that a propagating mode exists below the plasma frequency even for arbitrarily small hole size [Shin et al., 2005, Webb and Li, 2006]. The influence of this mode on the transmission properties of sub-wavelength circular apertures has yet to be clarified experimentally.

A second important point we have to consider is that the spectral position of the localized surface plasmon mode will depend on the dimensions and geometrical form of the aperture. By analogy to the discussion of localized modes in metal nanoparticles and nanovoids in chapter 5, a significant field-enhancement at the aperture rim can be expected, which will increase the transmission at the wavelength where the localized mode is excited. It is only recently that advancements in the fabrication of single holes in free-standing metal films using focused ion beam milling have enabled careful studies of this phenomenon. Using this technique, Degiron and co-workers confirmed the signature of a localized plasmon mode in a single circular hole in a free-standing silver film (Fig. 8.11a) [Degiron et al., 2004]. For a relatively thin yet opaque metal film where appreciable tunneling through the aperture can take place, a peak in transmission was observed (Fig. 8.11b), and attributed to the excitation of a localized mode. Furthermore, the spatial structure and spectral signature

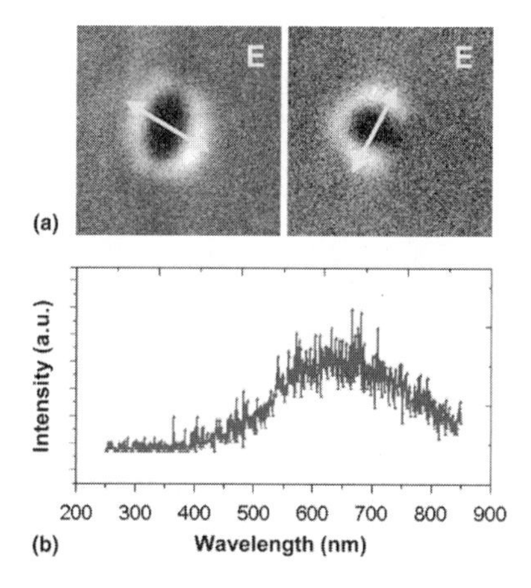

Figure 8.12. Electron-beam induced surface plasmon emission of light. (a) Cathodoluminescence image for two different polarizations. (b) Corresponding spectrum. Reprinted from [Degiron et al., 2004], copyright 2004, with permission from Elsevier.

of the localized plasmon mode could be established using excitation with a high-energy electron beam. Fig. 8.12 shows the beam-induced light emission (a) and corresponding spectrum (b), which shows good agreement with the spectral dependency $T(\lambda)$. Furthermore, the same study also presented first evidence of narrow beaming-effects due to a localized mode at the exit side of the screen.

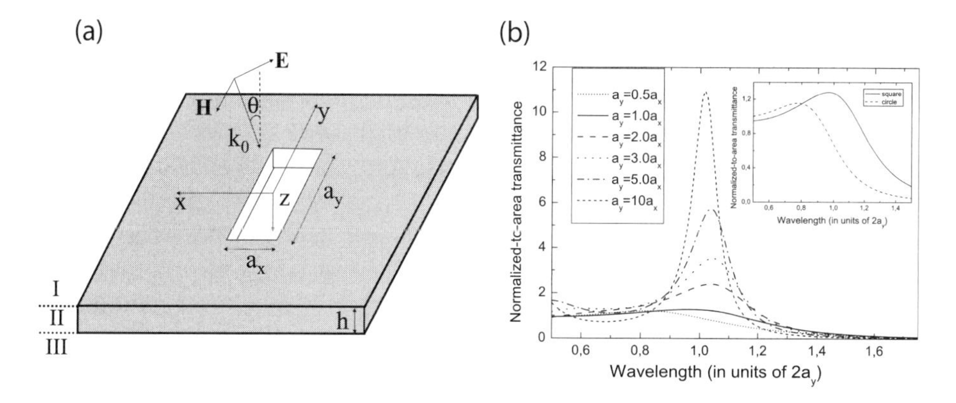

Figure 8.13. Transmission through a single rectangular aperture in a perfectly conducting screen. (a) Sketch of the geometry. (b) Normalized transmittance T versus wavelength for a normal incident plane wave impinging on apertures of different aspect ratio a_y/a_x. The thickness of the metal is $h = a_y/3$. The inset compares the transmission between a single square and circular aperture. Courtesy of Francisco García-Vidal, Universidad Autónoma de Madrid. Reprinted with permission from [García-Vidal et al., 2005b]. Copyright 2005 by the American Physical Society.

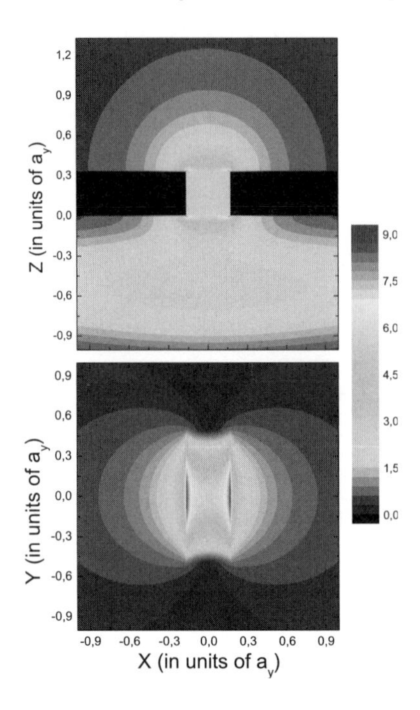

Figure 8.14. Enhancement of the electric field $|\mathbf{E}|$ with respect to the incident field for a rectangular aperture of Fig. 8.13 with $a_y/a_x = 3$ and $h = a_y/3$ at the resonant wavelength. The top panel shows a cut through the center of the aperture, and the lower panel the field distribution at the entrance surface. Reprinted with permission from [García-Vidal et al., 2005b]. Copyright 2005 by the American Physical Society.

A recent study has suggested that localized modes also play a role in the transmission through periodic arrays of sub-wavelength apertures [Degiron and Ebbesen, 2005]; however, compared to the importance of propagating SPPs discussed above, the localized modes incur only minor changes [de Abajo et al., 2006, Chang et al., 2005].

A related work by García-Vidal and colleagues analyzed the transmission resonances occurring for a rectangular aperture of varying aspect ratios a_y/a_x as depicted in Fig. 8.13a [García-Vidal et al., 2005b]. In an important difference to the experimental work [Degiron et al., 2004], the metal screen was modeled as a perfect conductor. Thus, excitation of a localized surface plasmon mode was excluded by the boundary conditions along the rim of the hole, just as in our discussion of perfect conductor modeling in the low frequency regime presented in chapter 6. A modal analysis of the fields in the half spaces above and below the screen, as well as in the aperture region of depth h, revealed a resonance in $T(\lambda)$ (Fig. 8.13b) near cut-off that increased in strength with a_y/a_x and the amount of dielectric filling of the hole. As in the optical study, this enhancement is due to a resonance, depicted in Fig. 8.14, which is however not of the nature of a surface plasmon. The rich physics of the transmission process in the cross-over regime between decaying and propagating modes has been revealed in a similar study [Bravo-Abad et al., 2004b].

To conclude this section, we want to point out that field tunneling through a single aperture mediated via SPPs can be enhanced by strengthening the coupling between the input and exit interfaces, for example by the introduction of a multilayer structure into the metallic screen [Chan et al., 2006, Zayats and Smolyaninov, 2006], or by filling the hole with a high-index dielectric [Olkkonen et al., 2005].

8.5 Emerging Applications of Extraordinary Transmission

Frequency-selective enhanced light transmission (through aperture arrays and even single apertures mediated by SPPs, localized surface plasmons or aperture waveguide resonances) is of course not only intriguing from a fundamental standpoint, but also for use in practical applications. A number of theoretical and recently also experimental studies have exploited the associated heightened fields at the maxima of $T(\lambda)$ for applications in optical switching using a suitable non-linear filling material [Porto et al., 2004], or for the enhancement of fluorescent emission from molecules located inside the aperture [Rigneault et al., 2005]. The goal of the nonlinear work is the demonstration of all-optical, electrical or thermal switching of the transmission. The physics of emission enhancement in the near-field of metallic structures will be covered in chapter 9.

A boost in the light transmission through nanoscale apertures is further of immediate interest for applications in near-field optics. While the Bethe-Bouwkamp treatment has recently been adapted to the conical geomtery of a typical near-field optical probe [Drezet et al., 2001], it is up to now not clear how lessons learned for a planar geometry can be applied to the design of more efficient near-field probes.

Schouten and co-workers have recently demonstrated the consequences of plasmon-assisted transmission on the classic Young's experiment of diffraction of light by a double slit [Schouten et al., 2005]. Another noteworthy extension of the principles presented here is the prediction of resonant transmission of cold atoms through sub-wavelength apertures in a screen sustaining matter waves [Moreno et al., 2005].

8.6 Transmission of Light Through a Film Without Apertures

At the end of this chapter, we want to briefly touch on the subject of light transmission mediated by surface plasmons through a metal film *without* holes. Chapter 2 described how for a metal film of a thickness smaller than the skin depth, interaction between SPPs sustained at the top and bottom interface takes place, leading to the establishment of coupled bound and leaky SPP modes. For infinitely wide thin metal layers embedded in a symmetric dielectric host, the two bound coupled modes are of distinct parities and have the opposite behavior regarding field confinement with vanishing layer thickness. If both surfaces are additionally modulated via a grating structure, SPPs can be excited on one side of the interface by direct light illumination via grating coupling, tunnel through the film, and be re-emitted on the other side if the period and height of the gratings on both sides are equal. This form of light transmission through a corrugated, unperforated metal film is accompanied by strong energy localization in the grooves of the grating at the input and the exit side [Tan et al., 2000].

While we might naively expect that the transmission efficiency monotonically increases with decreasing thickness of the metal film due to the increase in overlap between the SPP modes at the two interfaces, for metal films situated on a high-index substrate such as a prism the transmission coefficient can in fact show a maximum for a certain film thickness d_{crit}. This is due to the competing effects of increased absorption and but also increased optical field enhancement with increasing d: the reduction in leakage radiation into the prism more than offsets the increase in absorption, which was demonstrated using direct illumination with grating coupling [Giannattasio et al., 2004], and local excitation using a superstrate doped with fluorescent dyes [Winter and Barnes, 2006].

Hooper and Sambles demonstrated that a rich new physics evolves if the gratings on both sides of the film are dissimilar [Hooper and Sambles, 2004a]. For certain conditions, extraordinary transmission similar to that occurring in metal films perforated with apertures can take place, and applications to enhance the external quantum efficiency of (for example) organic light emitting diodes have been suggested [Wedge et al., 2004]. Similar phenomena were shown to appear in two-dimensionally corrugated metal films [Bonod et al., 2003, Bai et al., 2005].

All these studies have effectively focused on coupled SPP modes in dielectric/metal/dielectric three-layer structures. Enhanced transmission can also take place via the bound mode in the opposite metal/dielectric/metal structure, where a highly localized mode is excited in the gap between two metal surfaces. A recent study using near-field imaging has provided first confirmation of these effects [Bakker et al., 2004]. We will return to the discussion of light transmission through a flat metal film in a context of imaging in chapter 11.

Chapter 9

ENHANCEMENT OF EMISSIVE PROCESSES AND NONLINEARITIES

One of the most spectacular applications of plasmonics to date is *surface enhanced Raman scattering* (SERS), which exploits the generation of highly localized light fields in the near-field of metallic nanostructures for enhancing spontaneous Raman scattering of suitable molecules. Using chemically roughened silver surfaces, Raman scattering events of single molecules have been recorded [Kneipp et al., 1997, Nie and Emery, 1997], with estimated enhancements of the scattering cross section by factors up to 10^{14}. The majority of this enhancement is believed to arise from the highly enhanced fields in metal nanoparticle junctions due to localized surface plasmon resonances. Termed *hot spots*, these highly confined fields also enable an increase of fluorescent emission, albeit with more modest enhancement factors. A proper understanding and control over the generation of these hot spots, for example in the form of nanoscale plasmonic cavities, is currently one of the major driving forces behind the design of nanoparticle ensembles with tuned optical properties.

This chapter will focus mainly on the fundamentals and geometries for SERS due to localized plasmon modes in metal nanostructures. Theoretical modeling based on scattering-type calculations will be reviewed, and additionally a cavity model for SERS presented, which aims to provide a general design principle and scaling law for this light-matter interaction. The related enhancement of fluorescence from emitters placed into the near-field of metallic nanostructures, as well as quenching processes due to non-radiative transitions, are treated as well. Enhancement of the intrinsic luminescence of noble metal nanoparticles and nonlinear processes are discussed at the end of this chapter.

9.1 SERS Fundamentals

The Raman effect (in the context of molecules) describes the inelastic scattering process between a photon and a molecule, mediated by a fundamental vibrational or rotational mode of the latter, as depicted in Fig. 9.1a. Due to energy exchange between the scattering partners, the incoming photon of energy $h\nu_L$ is shifted in energy by the characteristic energy of vibration $h\nu_M$. These shifts can be in both directions, depending on whether the molecule in question is in its vibrational ground state or in an excited state. In the first case, the photon loses energy by excitation of a vibrational mode (Stokes scattering). In the second case, additionally energy gains by de-excitation of such a mode (anti-Stokes scattering) are possible. The frequencies of these two Raman bands are therefore

$$\nu_S = \nu_L - \nu_M \tag{9.1a}$$

$$\nu_{aS} = \nu_L + \nu_M. \tag{9.1b}$$

Fig. 9.1b shows a comparison between a typical fluorescence and a Raman spectrum. As can be seen, whereas the former spectrum is usually relatively broad due to nonelastic electron relaxation to the lower edge of the excited level (see schematic), Raman transitions are much sharper, thus enabling a detailed analysis of the molecule under study. In general, the photons involved in Raman transitions are not in resonance with the molecule, and the excitation takes place via virtual levels. No absorption or emission of photons is involved, and the transition is a pure scattering process. This is true even in the case where the incoming photon is in resonance with an electronic transition. This resonant Raman scattering is stronger than normal Raman scattering, but its efficiency is still much weaker than that of fluorescent transitions. Typical Raman scattering cross sections σ_{RS} are usually more than ten orders of magnitude smaller than those of a fluorescent process: 10^{-31} cm^2/molecule $\leq \sigma_{\mathrm{RS}} \leq 10^{-29}$ cm^2/molecule, depending on whether the scattering is non-resonant or resonant.

The Raman scattering described here is a *spontaneous* (as opposed to stimulated) scattering event and thus a linear process: The total power of the inelastically scattered beam scales linearly with the intensity of the incoming excitation beam. We will in the following discuss the Stokes process, for which the power of the scattered beam can be expressed as

$$P_S(\nu_S) = N\sigma_{RS} I(\nu_L), \tag{9.2}$$

where N is the number of Stokes-active scatterers within the excitation spot, σ_{RS} is the scattering cross section, and $I(\nu_L)$ the intensity of the excitation beam.

SERS describes the enhancement of this process, accomplished by placing the Raman-active molecules within the near-field of a metallic nanostructure. The nanostructure can consist of metal colloids, specifically designed nanoparticle ensembles, or the topography of a roughened surface. The enhancement

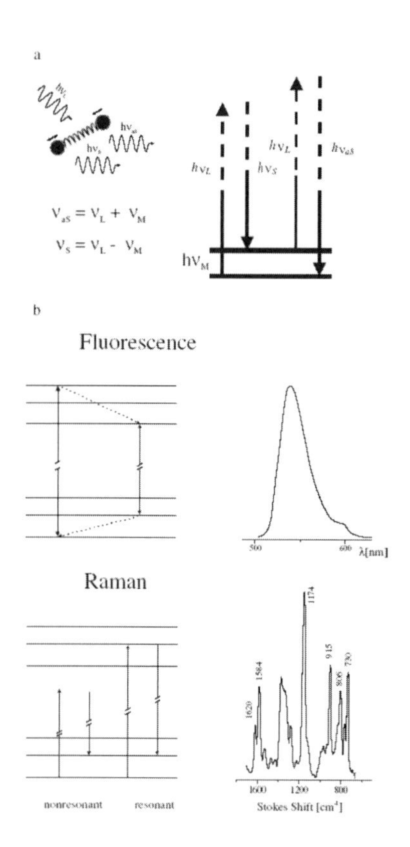

Figure 9.1. Schematic depiction of Raman scattering and fluorescence. (a) Generation of Stokes and anti-Stokes radiation via scattering events. (b) Fluorescence and Raman scattering in energy level pictures and representative spectra. Reprinted with permission from [Kneipp et al., 2002]. Copyright 2002, Institute of Physics.

of P_S is due to two effects. Firstly, the Raman cross section σ_{RS} is modified, due to a change in environment of the molecule. This change to $\sigma_{SERS} > \sigma_{RS}$ is often called the *chemical* or *electronic* contribution to the Raman enhancement. Theoretical modeling suggests that the maximum enhancement due to the change in cross section is of the order of 100.

A much more important factor in the total enhancement of P_S is the increased electromagnetic field due to excitations of *localized surface plasmons* and a crowding of the electric field lines (*lightning rod effect*) at the metal interface [Kerker et al., 1980, Gersten and Nitzan, 1980, Weitz et al., 1983]. This leads to an enhancement of both the incoming and emitted light fields, expressed via $L(\nu) = |\mathbf{E}_{\mathbf{loc}}(\nu)| / |\mathbf{E}_0|$, where $|\mathbf{E}_{\mathrm{loc}}|$ is the local field amplitude at the Raman active site. $L(\nu)$ is called the *electromagnetic* enhancement factor. The total power of the Stokes beam under SERS conditions is

$$P_S(\nu_S) = N\sigma_{SERS} L(\nu_L)^2 L(\nu_S)^2 I(\nu_L). \tag{9.3}$$

Since the difference in frequency $\Delta\nu = \nu_L - \nu_S$ between the incoming and scattered photons is in general much smaller than the linewidth Γ of a localized surface plasmon mode, $|L(\nu_L)| \approx |L(\nu_S)|$, which brings us to the important result that the *electromagnetic contribution to the total SERS enhancement is proportional to the fourth power of the field enhancement factor.* The commonly used expression [Kerker et al., 1980] for the enhancement of the power of the Stokes beam is therefore

$$R = \frac{|\mathbf{E}_{\mathrm{loc}}|^4}{|\mathbf{E}_0|^4}. \tag{9.4}$$

We will not expand upon this elementary discussion of SERS, but concentrate on the field enhancement factor $L(\nu)$. The reader interested in a more detailed and rigorous discussion of SERS is instead referred to appropriate review articles [Kneipp et al., 2002, Moskovits, 1985].

The physical basis of the electromagnetic enhancement consists of two main contributions - the enhancement due to the resonant excitation of localized surface plasmons in metallic nanostructures, and the lightning rod effect [Gersten and Nitzan, 1980, Kerker et al., 1980, Liao and Wokaun, 1982]. Of the two phenomena, only the plasmon resonance shows a strong frequency dependence, while the lightning rod effect is due to the purely geometric phenomenon of field line crowding and the accompanying enhancement near sharp metallic features. Thus, we can write that $L(\nu) = L_{\mathrm{SP}}(\nu) L_{\mathrm{LR}}$. This description can be applied to both Raman, resonant-Raman, and fluorescent enhancement near metallic nanostructures.

The functional form of L_{SP} is essentially that of the polarizability α of the metallic nanostructure of a given geometrical shape. For a spherical nanoparticle of sub-wavelength diameter, we can thus write by recollecting (5.7)

$$L_{\mathrm{SP}}(\omega) \propto \frac{\varepsilon(\omega) - 1}{\varepsilon(\omega) + 2}. \tag{9.5}$$

Similarly, for ellipsoidal particles the appropriate form of the polarizability presented in chapter 5 has to be used, and L_{SP} describes then the field enhancement averaged over the particle surface. In this case, the additional field enhancement occurring at the tips of prolate ellipsoids due to the continuity of the dielectric displacement field is described via the lightning rod factor L_{LR}, scaling with the ratio of the permittivities of the metal and the surrounding dielectric (usually air). For more complex geometries, in general the enhancement factors have to be calculated numerically.

9.2 SERS in the Picture of Cavity Field Enhancement

A slightly different view of SERS describes the enhancement process via the interaction between the molecule and an electromagnetic cavity mode. This cavity can for example be formed by the junction between two closely spaced metal nanoparticles, which is believed to be the site for hot-spots in experiments where single-molecule SERS was observed [Kneipp et al., 1997, Nie and Emery, 1997]. The enhancement of the electromagnetic field in such a cavity can be expressed via its quality factor Q, describing the *spectral* mode energy density, and its effective mode volume V_{eff}, describing the *spatial* mode energy density. We have seen in chapter 2 that SPPs propagating in a gap between two closely spaced metallic surfaces can show an effective mode *length* smaller than the diffraction limit of the dielectric filling medium. The same is true for the effective mode *volume* in plasmon cavities composed of such structures, and for localized modes in metal nanoparticles.

Using the concept of waveguide-to-cavity coupling to analyze the enhancement of an incoming beam by a metallic nanostructure [Maier, 2006b], a spontaneous Raman scattering process can be described by an incoming excitation beam of intensity $|\mathbf{E}_i(\omega_0)|^2/2\eta$ (η is the impedance of free space) and frequency ω_0, exciting a Raman active molecule in a cavity to emit Stokes photons at frequency ω via a scattering event. As mentioned in the preceding chapter, due to the small Stokes emission shift, one can assume equal enhancement of the exciting field and the outgoing Stokes field. In a context of field enhancement in a cavity, we can therefore write $Q(\omega_0)$=$Q(\omega)$=Q and $V_{\text{eff}}(\omega_0)$=$V_{\text{eff}}(\omega)$=V_{eff}, assuming that both the incoming and the emitted photons are resonant with the cavity. In order to calculate the enhancement, we want to obtain an expression for R, defined via (9.4), in terms of Q and V_{eff}.

With $|s_+|^2 = |\mathbf{E}_i|^2 A_i/2\eta$ being the power carried by the incident beam of cross section A_i, the evolution of the *on-resonance* mode amplitude u inside the cavity can be calculated using the relation $\dot{u}(t) = -\frac{\gamma}{2}u(t) + \kappa s_+$ [Haus, 1984], where u^2 represents the total time averaged energy in the cavity. $\gamma = \gamma_{\text{rad}} + \gamma_{\text{abs}}$ is the energy decay rate due to radiation (γ_{rad}) and absorption (γ_{abs}), and κ is the coupling coefficient to the external input, which depends on the size and shape of the excitation beam. κ can be expressed as $\kappa = \sqrt{\gamma_i}$, where γ_i is the contribution of the excitation channel to the total radiative decay rate [Haus, 1984]. For a symmetric two-sided cavity, in a first approximation one can estimate $\gamma_i = (\gamma_{\text{rad}}/2)(A_c/A_i)$, with A_c corresponding to an effective radiation cross-section of the resonant cavity mode (its radiation field imaged back into the near-field of the cavity). Note that A_i has been assumed to be larger than A_c in the above relation, and that A_c can be no smaller than the diffraction limited area A_d ($A_d \leq A_c \leq A_i$). Putting everything together, in steady state the mode amplitude can be expressed as [Maier, 2006b]

$$u = \frac{\sqrt{2\gamma_{\text{rad}} A_c / A_i} |s_+|}{\gamma_{\text{rad}} + \gamma_{\text{abs}}} = \frac{\sqrt{\gamma_{\text{rad}} A_c} |\mathbf{E}_i|}{\sqrt{\eta}(\gamma_{\text{rad}} + \gamma_{\text{abs}})}, \tag{9.6}$$

which for fixed incoming power is maximum upon spatial mode matching ($A_c = A_i$).

Due to the different contributions of radiative and absorptive damping, we now have to distinguish between dielectric and metallic cavities. For a dielectric cavity ($\gamma_{\text{rad}} \gg \gamma_{\text{abs}}$), $u \propto 1/\sqrt{\gamma_{\text{rad}}} \propto \sqrt{Q}$, while for a metallic cavity ($\gamma_{\text{abs}} \gg \gamma_{\text{rad}}$) $u \propto 1/\gamma_{\text{abs}} \propto Q$, explaining the different scaling laws for field enhancement in dielectric [Spillane et al., 2002] and metallic [Klar et al., 1998] resonators encountered in the literature.

Since the effective mode volume relates the local field to the total electric field energy of the cavity (see the discussion of the effective mode length in chapter 2), we can write the resonant mode amplitude as $u = \sqrt{\varepsilon_0} |\mathbf{E}_{\text{loc}}| \sqrt{V_{\text{eff}}}$. Therefore, using (9.6) the enhancement of the incoming radiation in a metallic cavity evaluates to

$$\sqrt{R} = \frac{|\mathbf{E}_{\text{loc}}|^2}{|\mathbf{E}_i|^2} = \frac{\gamma_{\text{rad}} A_c}{4\pi^2 c^2 \eta \varepsilon_0 \lambda_0} \frac{Q^2}{\bar{V}_{\text{eff}}}. \tag{9.7}$$

A similar scaling law has been obtained for plasmonic energy localization in fractal-like metal nanoparticle aggregates on metal surfaces [Shubin et al., 1999].

We can now use this expression to estimate R for a crevice between two silver nanoparticles separated by a nanoscale gap, a configuration which is believed to sustain SERS hot-spots with $R \sim 10^{11}$ upon resonance. The crevice can be approximately modeled as a metal/air/metal heterostructure treated in chapter 2, with the lateral widths fulfilling a Fabry-Perot-like resonance condition: the fundamental resonance occurs when half the wavelength of the coupled SPP mode fits inside the cavity. Its effective dimensions are thus the effective mode length L_z of the gap structure, calculated using the procedure outlined in chapter 2, and $L_y \sim L_x = \lambda_{\text{SPP}}/2 = \pi/\beta$. Using the simplified analytical treatment of a one-dimensional silver/air/silver structure with a 1 nm air gap for the calculation of β and L_z, $A_c = A_d$, and (Q, γ_{rad}) estimated from FDTD calculations, (9.7) yields $R \sim 2.7 \times 10^{10}$ for excitation at $\lambda_0 = 400$nm, in good agreement with full-field three-dimensional simulations of the enhancement for this coupled particle geometry [Xu et al., 2000].

The total observable enhancement of the Stokes emission can be estimated as the product of the field enhancement of the incoming radiation and the enhanced radiative decay rate at the Stokes frequency. As is well known, a dipole oscillator placed inside a metallic cavity shows an increase in its *total* decay rate $\gamma/\gamma_0 = (3/4\pi^2)(Q/\bar{V}_{\text{eff}})$[Hinds, 1994].

However, we have to note that the dominance of absorption over radiation as loss channels has to be taken into account. For collection of light emission outside the cavity, the overall cavity enhancement must therefore be weighted with an extraction efficiency, Q/Q_{rad} [Barnes, 1999, Vuckovic et al., 2000]. The emission enhancement at the *peak* emission frequency of the Stokes line can then be written as $(3/4\pi^2)(Q^2/\bar{V}_{\mathrm{eff}})(Q/Q_{\mathrm{rad}})$. Incorporating the relation for the enhancement of the excitation field (9.7), the overall enhancement is estimated to be 1.5×10^{12} for the crevice example, similar to observed values [Nie and Emery, 1997, Kneipp et al., 1997]. More details on this model can be found in [Maier, 2006b].

9.3 SERS Geometries

In this section we will discuss a number of important geometries where large enhancements of Raman scattering have been experimentally observed. Since in order to achieve local field enhancement, a surface showing strong localized plasmons is desirable, ensembles of metallic nanostructures with interstitial gaps of the order of only a few nanometers are preferable. Furthermore, the intrinsic response of the metal, expressed via its dielectric function $\varepsilon(\omega)$, has to allow for the resonances to occur in the spectral regime of interest. Since up to now most studies have been limited to gold and silver, SERS with high enhancement factors is mostly reported for work in the visible regime of the spectrum.

As already mentioned at a number of occasions, the highest enhancements recorded to date have been achieved on roughened silver surfaces and are on the order of 10^{14} [Kneipp et al., 1997, Nie and Emery, 1997]. It is believed that the electromagnetic effect provides for a factor of up to 10^{12} to this total enhancement. Taking the scaling of the Raman enhancement with the local field amplitude (9.4) into account, the rough surface must therefore support hot spots with field enhancement factors $L(\nu)$ on the order of 1000.

García-Vidal and Pendry modeled this geometry as a collection of closely spaced semicylinders on a flat surface (Fig. 9.2, left) [García Vidal and Pendry, 1996]. The SERS enhancement provided by this topography was calculated using a scattering analysis, which yielded $R \approx 10^8$ for interstitial sites between touching cylinders (Fig. 9.2, right). The highly localized field at such a site is depicted in Fig. 9.3, and is seen to arise from a localized plasmon mode in the gap region between the two metallic surfaces. The conduction electrons in the two touching cylinders move as to create an opposite charge density distribution on neighboring surfaces; thus, the mode is related to the coupled SPP mode in a metal/air/metal heterostructure, described in chapter 2 and the preceding section. The importance of localized gap-modes for SERS was further corroborated in comprehensive numerical electromagnetic studies of interstitial sites between metal nanoparticles, which confirmed that enhancements

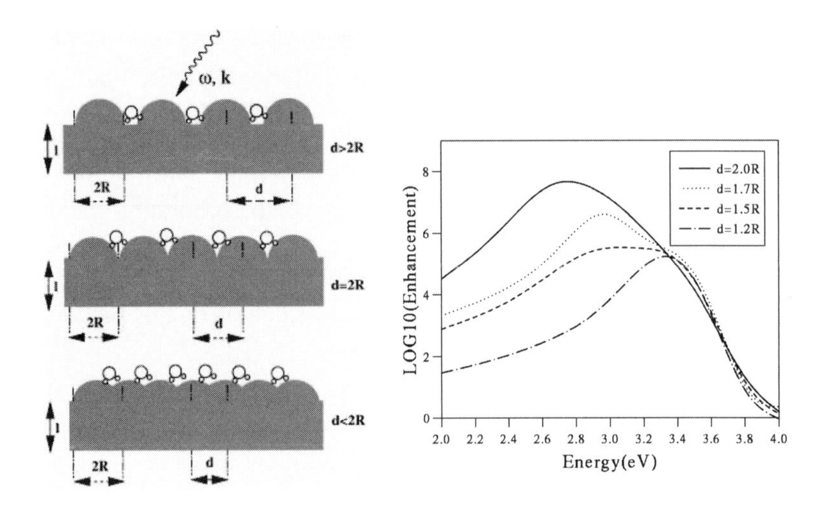

Figure 9.2. Sketch of a rough metal surface modeled via a chain of silver semicylinders placed upon a flat silver surface (left) and the local enhancement evaluated at the crevices between the semicylinders (right) for varying gap sizes. Reprinted with permission from [García-Vidal and Pendry, 1996]. Copyright 1996 by the American Physical Society.

enabling single-molecule detection are possible [Xu et al., 2000]. These investigations have also confirmed enhanced optical forces polarizing the molecules and attracting them into the gaps via the strong field gradient [Xu et al., 2002].

The realization that localized plasmons play a crucial role in the Raman enhancement of molecules at a metal surface has triggered a great amount of

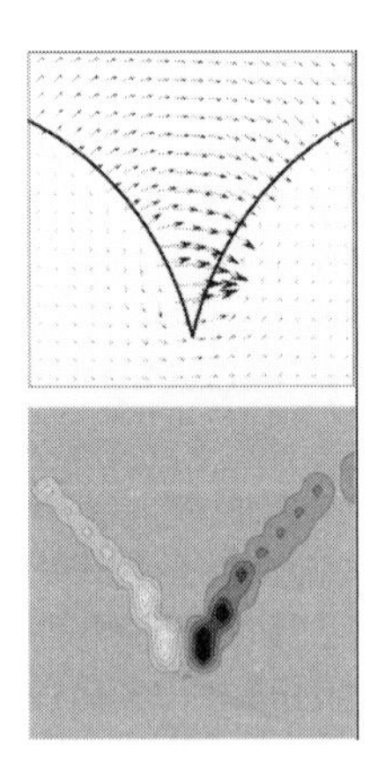

Figure 9.3. Distribution of the electric field (upper figure) and its divergence (lower figure) at the junction between two semicylinders for the geometry depicted in Fig. 9.2. Reprinted with permission from [García-Vidal and Pendry, 1996]. Copyright 1996 by the American Physical Society.

research into the design and fabrication of SERS substrates with controlled surface structure optimized for field enhancement. Topographies based on closely spaced nanoparticles (in a sense mimicking a surface with controlled, regular roughness), specially shaped nanostructures or nanovoids have been analyzed for their effectiveness as SERS substrates.

For example, SERS based on isolated metallic nanoparticles has been characterized using far-field Raman spectroscopy of regular particle arrays situated on a metal film substrate, where the localized surface plasmon resonance is mediated by far-field coupling between the particles [Félidj et al., 2004, Laurent et al., 2005a] as described in chapter 5. Studies of nanoparticles of various shapes have confirmed the crucial role of localized surface plasmon modes on the Raman enhancement [Grand et al., 2005], and multipolar excitations in elongated particles have also been shown to contribute to SERS [Laurent et al., 2005b]. Another promising particle geometry are metallic nanoshells [Xu, 2004, Talley et al., 2005], which can show large field-enhancements due to reduced plasmon linewidths at near-infrared frequencies. The field enhancement due to localized surface plasmon resonances can further be increased by placing the particles into a microcavity [Kim et al., 2005], or by coupling the localized plasmon to propagating SPPs on a continuous metal film [Daniels and Chumanov, 2005].

An example of a flat metal film structured with a nanovoid lattice is shown in Fig. 9.4 [Baumberg et al., 2005]. In this case, the voids support localized plasmon resonances and further act as a lattice for phase-matching for the excitation of SPPs (Fig. 9.4b). The plasmon is then Raman-scattered by the molecule into a plasmon of lower frequency, which is subsequently scattered into a photon. However, in their current form the electromagnetic field enhancement of

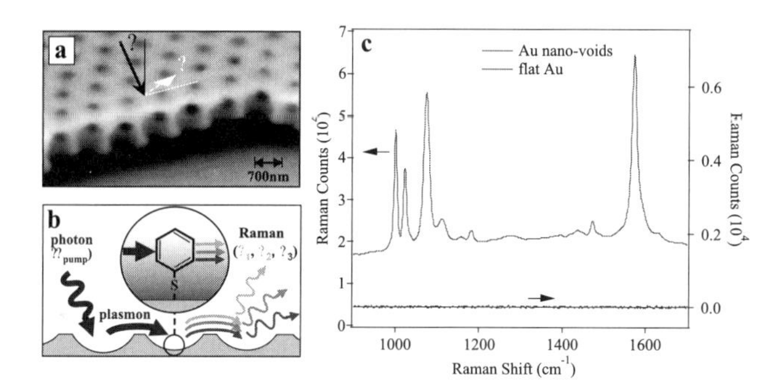

Figure 9.4. SERS using a nanovoid metal film. (a) SEM of the structured flat surface. (b) Schematic of the SERS process. (c) Example SERS spectrum. Reprinted with permission from [Baumberg et al., 2005]. Copyright 2005, American Chemical Society.

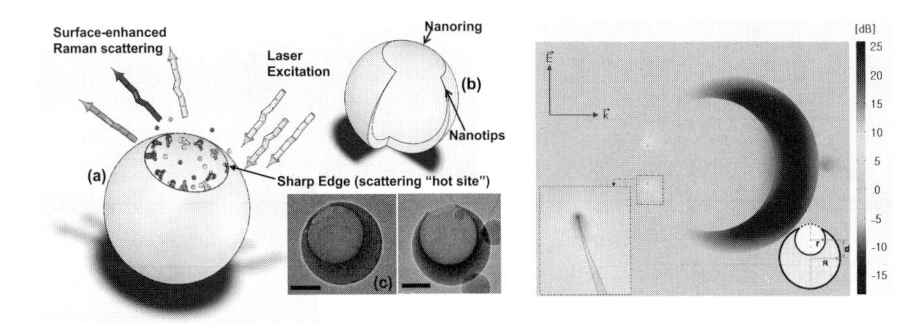

Figure 9.5. Fabrication process of crescent moon structures (left) and electric field profile (right) showing hot-spots at the tips of the moon structure. Reprinted with permission from [Lu et al., 2005]. Copyright 2005, American Chemical Society.

these nanovoid-decorated flat films is lower than that of rough surfaces where single-molecule Raman was observed.

In order to achieve an electromagnetic field enhancement of the order of 1000 necessary for single-molecule Raman with nanofabricated structures, nanometric gaps between metallic surfaces akin to those naturally occuring on rough surfaces have to be achieved. One strategy involves the fabrication of metal nanoparticles in the form of a crescent moon with two sharp tips spaced by only a small gap [Lu et al., 2005]. Lu and co-workers realized the fabrication of such particles via angled metallization of nanospheres (Fig. 9.5 left). Electromagnetic simulations show a high field-enhancement at the sharp tip (Fig. 9.5 right), which is believed to be due to localized plasmon resonances and the lightning rod effect. The field enhancement at the tips is in excess of 100, leading to a Stokes enhancement of the order of 10^{10}. Similar enhancements can be achieved in small gaps between opposing nanotriangles [Sundaramurthy et al., 2005].

Another promising geometry for reliable SERS substrates are aligned, high-aspect nanowires fabricated using a porous templating process. Fig. 9.6 shows an example of a SERS spectrum and SEM of a silver nanowire array fabricated using a porous alumina template [Sauer et al., 2005]. Also, the use of porous silicon as a substrate for the generation of dentritic metal structures has been demonstrated [Lin et al., 2004].

Most SERS studies using substrates with nanostructured topographies have focused on the metals gold and silver, which show a localized plasmon resonance in the visible or near-infrared regime (for elongated particles), and are thus suitable for Raman in this spectral region. In order to extend SERS into different frequency regimes, particularly the ultraviolet region, a number of different metals have recently started to be investigated, amongst them nickel [Sauer et al., 2006]. Additionally, rhodium and ruthenium seem to show partic-

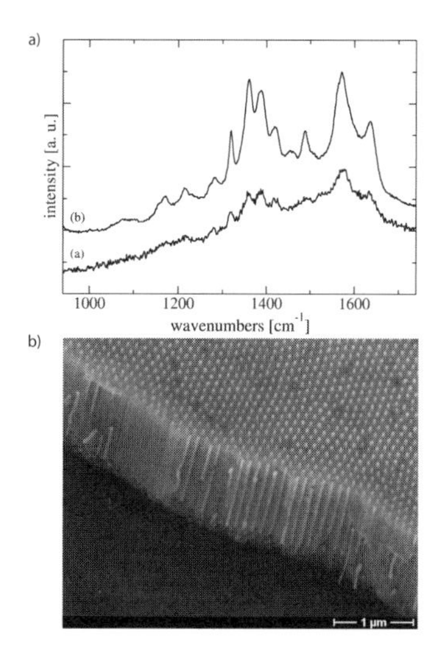

Figure 9.6. SERS spectra (a) and SEM image (b) of a gold nanowire array in a porous alumina matrix. Reprinted with permission from [Sauer et al., 2005]. Copyright 2005, American Institute of Physics.

ular promise for applications in the UV [Ren et al., 2003, Tian and Ren, 2004], albeit with modest enhancement factors.

While metallic surfaces with a topography suitable for SERS show a high promise as platforms for biological and chemical sensing, many applications (especially in materials science) use Raman scattering to investigate, not single molecules, but thin-film samples of semiconductors and adsorbed species. In this case, spatially resolved Raman spectra are desirable, which are usually generated by scanning the excitation beam over the film under study using an optical microscope. To enable the enhancement of the Raman signal using this geometry, *tip-enhanced* Raman scattering [Lu, 2005] is required. In this case, a sharp metal tip is scanned over the surface using feedback akin either to STM, AFM or tuning-fork feedback. The tip is illuminated from the outside via a focused laser beam, thus creating an enhanced field at its apex due to localized resonances and the lightning rod effect. In order to observe a high field enhancement at the apex of the tip, the illumination condition has to be chosen such as to create a longitudinal dipolar charge distribution. Using illumination from the bottom, this requires highly focused Gaussian beams [Hayazawa et al., 2004] or the use of Hermite-Gaussian beams, which show a strong longitudinal field component. We note that on metalized tips of conical shapes, field enhancement can arise both from localized modes at the (spherical) apex, as well as from surface modes sustained by the surface of cone. As an example, Fig. 9.7 shows the electric field enhancement at a metal tip calculated

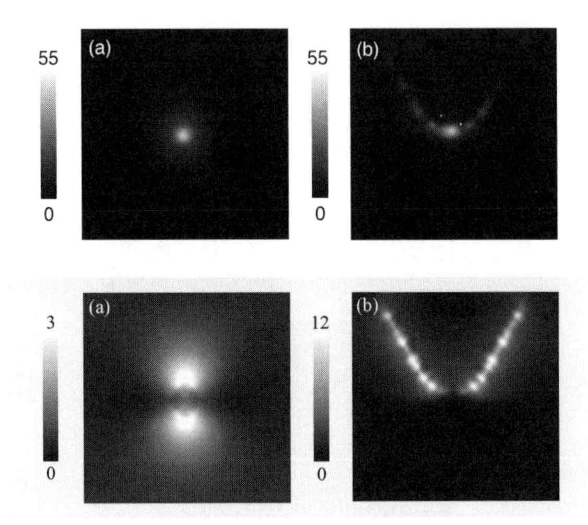

Figure 9.7. FDTD calculations of the electric field enhancement at a silver cone of semiangle 30° terminated with a spherical apex of radius 20 nm. The upper row shows the field distribution at the resonance frequency of the apex when the tip is situated 2 nm above a glass substrate. In the pictures in the lower row the tip is illuminated at the surface plasmon resonance frequency of the silver cone. (a) Frontal view from the glass substrate side. (b) Side-view cut through the symmetry plane of the cone. Reprinted with permission from [Milner and Richards, 2001]. Copyright 2001, Blackwell Publishing.

using finite-difference time-domain modeling for illumination at an angle at the frequency $\omega_p/\sqrt{3}$ of the localized apex mode (upper row), and for normal illumination at the frequency $\omega_p/\sqrt{2}$ of the surface plasmon mode of the cone surface (lower row) [Milner and Richards, 2001].

Apart from the intrinsic enhancement at the apex of a sharp metal tip discussed in the context of SPP focusing in chapter 7, it is currently believed that the enhanced field in the tip-sample cavity contributes to the observed enhancements. These techniques have for example been applied to investigations of nucleotides [Watanabe et al., 2004] and small carbon-based molecules [Pettinger et al., 2004]. Resolution on the order of 25 nm has been demonstrated for carbon nanotube substrates [Hartschuh et al., 2003].

9.4 Enhancement of Fluorescence

The heightened electromagnetic fields near metallic surfaces due to localized plasmon resonances and propagating SPPs also enhance the emission of fluorescent species placed in the near field. However, for molecules in contact with the metallic surface, care has to be taken in order not to quench the fluorescence via non-radiative transitions. Thus, for the observation of enhanced fluorescence, often a nanometer-thin dielectric spacer layer is required to prohibit non-radiative excitation transfer from the molecule to the metal. We have already hinted at this point in chapter 4 when discussing fluorescence imaging of SPP propagation.

Let us briefly illustrate the complexity of the interaction process by focusing on one particular investigation. Anger and co-workers performed a comprehensive study of the enhancement and quenching of emission from a single fluorescent molecule near a sub-wavelength gold sphere [Anger et al., 2006]. Fluorescence results from excitation of the molecule by the incident field - which can show significant enhancement due to a plasmon resonance of the gold particle - and the subsequent emission of radiation by the molecule, which is determined by the balance between radiative and non-radiative decay processes. Since non-radiative energy transfer to the nanoparticle can take place for small distances between the molecule and the sphere, a decrease in emission probability can be expected, despite an increase in excitation rate due to the enhanced local field.

For weak excitation, the fluorescence emission rate γ_{em} can be related to the excitation rate γ_{exc} and the total decay rate $\gamma = \gamma_{\mathrm{r}} + \gamma_{\mathrm{nr}}$ via

$$\gamma_{\mathrm{em}} = \gamma_{\mathrm{exc}} \frac{\gamma_{\mathrm{r}}}{\gamma}, \tag{9.8}$$

where γ_{r} is the radiative and γ_{nr} the non-radiative decay rate. The emission probability $q_{\mathrm{a}} = \gamma_{\mathrm{r}}/\gamma$ is also called the quantum yield of the emission process. The fluorescence process can then be treated by assuming a two-level model of the molecular transition, and a description of the modified electromagnetic environment due to the presence of the gold nanoparticle using a Green's function approach. In this study, a profound difference was found for short separations z between the emitter and the sphere between treatments of the particle as a simple dipole, and a description involving multipolar orders. Fig. 9.8 shows results for the quantum yield q_a and the normalized excitation and fluorecent emission rates γ_{exc} and γ_{em} as a function of the distance between the mole cule and gold spheres of different sub-wavelength sizes. Taking into account higher order interactions (apart from simple dipolar coupling) confirms the experimentally observed emission quenching for small gaps between the emitter and the metallic structure, due to non-radiative energy transfer (Fig. 9.8b). It is interesting to note that since γ_{nr} is proportional to the amount of Ohmic heating, the maximum in fluorescence enhancement does not necessarily occur for excitation at the plasmon resonance frequency.

An experimental setup suitable for the observation of the predicted distance dependence of the fluorescent emission is shown in Fig. 9.9a. The gold sphere is attached to the scanning tip of a near-field optical microscope to allow the controlled variation of the distance to the molecule, which is placed on a planar

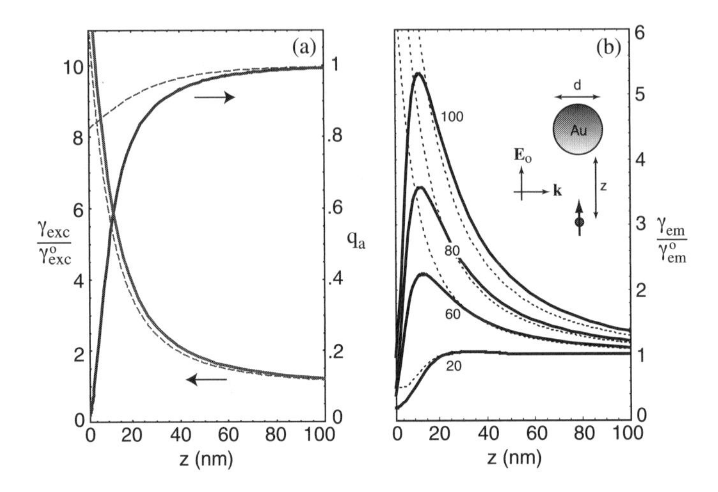

Figure 9.8. Calculated quantum yield q_a, excitation rate γ_{exc} and fluorescence rate γ_{em} for a single molecule located in a distance z from a gold sphere of diameter 80 nm (a) or as indicated in figure (b). Excitation takes place at $\lambda = 650$ nm, which was assumed to coincide with the peak of the emission spectrum, and all rates are normalized to their respective free-space values. The dashed lines correspond to a dipole model of the particle, and the continuous lines to a model taking higher multipoles into account. Reprinted with permission from [Anger et al., 2006]. Copyright 2006 by the American Physical Society.

substrate. Fig. 9.9b shows the calculated field distribution in the sphere-surface cavity.

A study of the single molecule emission rate versus vertical position of the tip revealed a functional dependence in agreement with the theoretical pre-

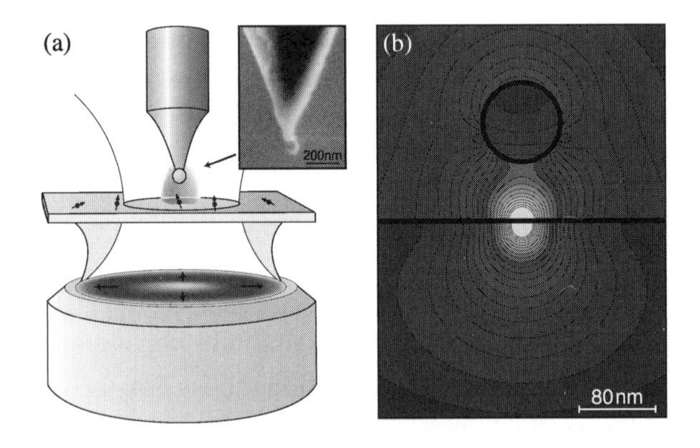

Figure 9.9. Experimental setup (a) and calculated field distribution for an emitter located on a glass substrate at a distance $z = 60$ nm below a gold particle (b) for the study of single-molecule fluorescence near a gold sphere. Reprinted with permission from [Anger et al., 2006]. Copyright 2006 by the American Physical Society.

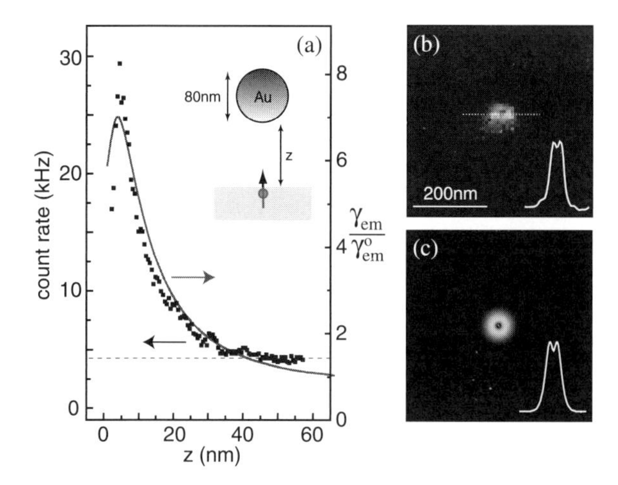

Figure 9.10. Experimentally determined emission rate (dots) and comparison with the theoretical curve of Fig. 9.8b) (a) and near-field image (b) of a fluorescent molecule near a gold sphere. A theoretical calculation of the emission intensity is shown in (c). Reprinted with permission from [Anger et al., 2006]. Copyright 2006 by the American Physical Society.

diction (Fig. 9.10a). Experimentally observed and calculated pictures of the single molecule emission are shown in panels b) and c) of this figure, and are in good agreement with each other. It is interesting to note that the decrease in quantum yield has not only been attributed to an increase in the non-radiative decay rate, but also to phase-induced decreases of the radiative decay process for small emitter-particle separations [Dulkeith et al., 2002]. While near-field optical microscopy is a convenient means to investigate the enhancement and quenching of fluorescent emission in a controlled fashion, also other promising geometries are emerging, such as for example metallic tunnel junctions filled with an organic layer with embedded molecules [Liu et al., 2006].

Xu and co-workers have shown that the enhancement of Raman scattering and fluorescence near a metallic surface or nanoparticle can be described using a unified treatment [Xu et al., 2004, Johannsson et al., 2005]. We will not carry the description of fluorescence further, but instead use the remainder of this chapter for a brief look at the enhancement of other emissive processes.

9.5 Luminescence of Metal Nanostructures

Photoluminescence from bulk noble metal samples was first observed by Mooradian using gold and copper samples excited by a strong (2 W) cw argon-ion laser beam [Mooradian, 1969]. The luminescence is due to the excitation of d-electrons into the sp-conduction band and subsequent direct radiative recombination, resulting in the peak of the luminescence spectrum being centered around the interband absorption edge. However, due to the dominance of nonradiative relaxation processes, the quantum efficiency of this process is very low, on the order of 10^{-10} for smooth metallic films.

Significant enhancements of the photoluminescence yield (up to 10^6) have been achieved using rough metal films [Boyd, 2003] and metallic nanoparticles [Link and El-Sayed, 2000, Wilcoxon and Martin, 1998, Dulkeith et al., 2004], akin to similar enhancements observed for Raman scattering. The enhancement can be explained with the model of enhanced localized fields due to plasmon excitation and the lightning rod effect, using the enhancement factor $L(\nu)$ introduced at the beginning of this chapter. Following the argument leading to the scaling of the Raman enhancement (9.3), the increase of photoluminescence in the local field model is expected to scale as

$$P_{\text{lum}} \propto L(\omega_{\text{exc}})^2 L(\omega_{\text{em}})^2, \tag{9.9}$$

where ω_{exc} and ω_{em} are the frequency of excitation and emission, respectively. This model naturally explains the observation that the broad luminescence band is significantly enhanced only at the spectral position of the spectrally sharper plasmon resonance, as confirmed by Link and co-workers by studying gold nanorods of different aspect ratios [Link and El-Sayed, 2000].

In the local field picture, the photoluminescence process is not inherently altered from that on flat surfaces, in the sense that light emission is caused by direct recombination between the sp and d bands, albeit in heightened lo-

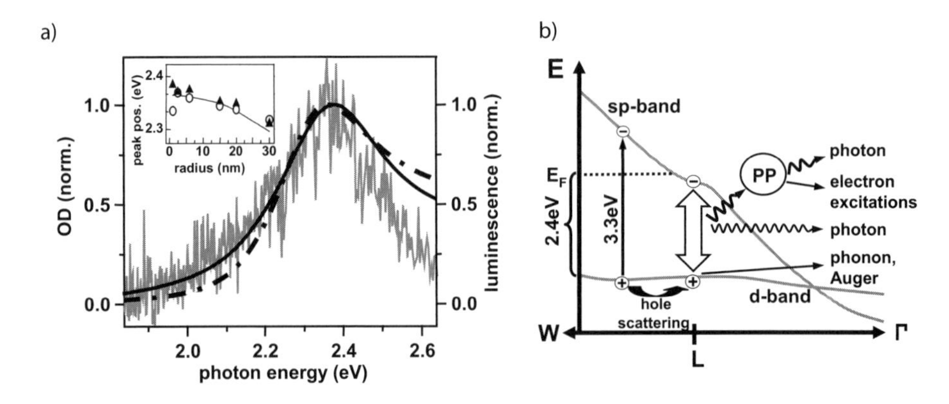

Figure 9.11. (a) Optical density (black line) and photoluminescence spectrum (gray line) for gold nanoparticles of radius 6 nm. The dashed-dotted line shows an extinction spectrum calculated using Mie theory. The inset shows the peak position of the optical density (triangles) and photoluminescence spectra (circles) for gold nanoparticles of different radii in solution. (b) Schematic of the plasmon-mediated photoluminescence process. After the initial excitation, the holes in the d-band may either radiatively recombine with electrons in the sp band, or non-radiatively via the creation of a particle plasmon, which decays either radiatively on non-radiatively. Reprinted with permission from [Dulkeith et al., 2004]. Copyright 2004 by the American Physical Society.

cal fields. A different model for the enhancement process was recently proposed by Dulkeith and co-workers in a study of the photoluminescence of gold nanospheres [Dulkeith et al., 2004]. As in the earlier studies, their observed luminescence spectrum closely followed that of the localized plasmon mode of the nanospheres (Fig. 9.11a). However, the obtained efficiency of 10^{-6} could not be explained using the local field model. Instead, a different model was proposed whereby a significant portion of the excited sp electrons decay into plasmons (Fig. 9.11b). The dominance of the plasmon decay channel was attributed to the large polarizability of the particle plasmon mode, leading to a greater radiative decay rate than that of a direct interband recombination. In this picture, the radiative decay of the plasmon into photons gives rise to the observed photoluminescence enhancement.

The luminescence processes discussed above are inherently linear or one-photon processes. Significant enhancement can also be achieved by using multi-photon absorption [Farrer et al., 2005], a description of which is however outside the scope of this book.

9.6 Enhancement of Nonlinear Processes

We want to conclude this chapter by presenting another category of emissive processes enhanced due to plasmonic field localization, namely that of nonlinear light generation. After the discussions above, it should come as no surprise that also nonlinear processes such as second or third harmonic generation can be strongly enhanced due to localized surface plasmons, as described by the local field model.

In principle, two different configurations exist, depending on whether the nonlinear effects are due to the intrinsic nonlinear susceptibility of the metal nanostructure itself, or caused by a nonlinear surrounding host. Both nonlinear processes are enhanced at frequencies within the lineshape of the localized plasmon. We will focus here on a brief description of the former process in the form of second harmonic generation from the metal nanostructures themselves.

The fact that metallic surfaces can emit second harmonic radiation in reflection despite the cubic symmetry of the metallic lattice is due to the breaking of the symmetry at the surface [Bloembergen et al., 1968, Rudnick and Stern, 1971, Sipe et al., 1980]. This process can be enhanced by the coupling to surface plasmons on flat films [Simon et al., 1974] or on films inscribed with a grating [Coutaz et al., 1985]. In the latter study, an enhancement factor of 36 compared with the flat film case was found. As with luminescence, significant enhancement of the second harmonic radiation can also be observed on rough metal surfaces [Chen et al., 1983], explained by the local field model. In this case, we expect the power P_{SH} of the second harmonic radiation to scale as

$$P_{\mathrm{SH}} \propto |L(2\omega)|^2 \left|L^2(\omega)\right|^2 . \tag{9.10}$$

For the calculation of the enhancement of a general, n-th order nonlinear process, one must simply replace each field $E(\omega)$ in the calculation of the nonlinear polarization by the local field $E_{\mathrm{loc}}(\omega) = L(\omega)\, E(\omega)$, with an additional enhancement factor for the emitted radiation at frequency $n\omega$.

Rough metal films can be viewed as a specific example of the more general case of composite optical materials with inherent randomness. The nonlinear optical properties of these small-particle composites are generally described within the framework of the *Maxwell-Garnett* model [Shalaev et al., 1996, Sipe et al., 1980]. A discussion of this theory is beyond the bounds of this book, but the interested reader is referred to the text by Shalaev on this topic [Shalaev, 2000].

Detailed studies of the enhanced second harmonic generation from rough metal films using laser scanning optical microscopy have revealed that the origin of the enhancement is indeed due to hot spots with high fields caused by localized modes [Bozhevolnyi et al., 2003]. For metal nanoparticles, detailed studies of the nonlinear properties have revealed important information about the inherent plasmon lifetime [Heilweil and Hochstrasser, 1985, Lamprecht et al., 1999] and the susceptibilities [Antoine et al., 1997, Ganeev et al., 2004, Lippitz et al., 2005].

Chapter 10

SPECTROSCOPY AND SENSING

The main part of this chapter describes different techniques for spectroscopic investigations of localized plasmon resonances in single metal nanoparticles, with a view to applications in sensing. The basic principle of single-particle sensors is the exploitation of the fact that the spectral position of their resonances depends on the dielectric environment within the electromagnetic near field. Applied to biological sensing, adsorption of molecules on a functionalized metal surface leads to spectral changes of the sustained plasmon modes. Due to the very localized nature and therefore high energy concentration in the near field of surface plasmons, even molecular monolayers can lead to discernible spectral changes. This high sensitivity has allowed surface plasmon sensors to become established as an analytical sensing technology over the last two decades.

The most important challenge encountered in almost any biosensor design is that of ensuring selectivity. In the case of surface-plasmon-based sensors, this is achieved via functionalization of the metallic surface to ensure only selective binding of the agent to be sensed. We will not focus on this aspect of sensor design here, but only mention that the surface chemistry of gold deserves special attention due to the relative ease of establishing sulfur bonds between gold atoms and organic molecules. Therefore, gold has emerged as the metal of choice for almost all practical optical sensing applications, including those based on surface plasmons. An important consequence is that due to the permittivity of gold, sensing is usually limited to the visible and near-infrared part of the spectrum.

We provide an overview of different excitation geometries for the investigation of localized surface plasmons, which is related to the analogous discussion of SPP excitation in chapter 3. The second part of this chapter aimes to give a flavor for different aspects of sensors based on propagating SPPs, relying on changes to the dispersion relation and the condition of phase-matching upon refractive index changes at flat metal interfaces. We will limit the discussion to two prominent excitation geometries, based on prism coupling and coupling using optical fibers coated with a metal film. However, we will not embark on a discussion of sensor performance in terms of selectivity and sensitivity. As a starting point for the exploration of important omitted aspects such as these, the reader is referred to the review by Homola [Homola et al., 1999].

10.1 Single-Particle Spectroscopy

This section continues the discussion of excitation mechanisms, presented in chapter 3 for propagating SPPs, by describing different excitation pathways for localized plasmon resonances in metal nanoparticles. We have seen in the description of the fundamentals of localized resonances in chapter 5 that the frequencies of the resonant modes of regular *particle ensembles* can be determined using conventional far-field extinction spectroscopy. Upon resonance the extinction cross sections of individual particles are resonantly enhanced, and for a sufficient spacing the extinction peak of the ensembles coincides with the localized plasmon frequency of an individual particle. However, due to slight differences in particle shape, inhomogeneous broadening of the extinction line shape can occur. Spectroscopy of *single nanoparticles* requires more sensitive detection techniques (due to the large background of radiation directly passing from the source to the detector), which will be outlined in this section.

The investigation of plasmon resonances of single particles is not only relevant from a fundamental point of view (e.g. the determination of the homogeneous linewidth Γ), but also potentially for practical applications in sensing. In this context, single metal nanoparticle sensors operate via the detection of frequency shifts of the dipolar plasmon resonance upon binding of molecules to the nanoparticle surface, which can be detected using spectroscopic techniques suitable for single-particle investigation.

Let us briefly review: for a spherical particle of a sub-wavelength diameter $d \ll \lambda_0$, the resonance frequency of the dipole mode for small damping is given by the Fröhlich condition

$$\varepsilon\left(\omega_{\mathrm{sp}}\right) = -2\varepsilon_{\mathrm{m}}. \tag{10.1}$$

Here, $\varepsilon(\omega)$ is the dielectric function of the metal, and ε_m the dielectric constant of the insulating host. Whereas the derivation of (10.1) in chapter 5 has assumed an infinite extent of the surrounding host medium, the sub-wavelength localization of the dipolar plasmon mode means that ω_{sp} is only determined by the dielectric environment within the tail of the evanescent near field of the particle. Changes in ε_{m}, induced for example via adsorption of a molecular monolayer on the particle surface, can then be detected via changes in the dipolar resonance frequency ω_{sp}.

While sensing in this manner can be easily performed using far-field extinction spectroscopy if a large amount of particles arranged in a regular array is used as the sensing template, sensors based on a *single* metallic nanostructure are highly desirable. Firstly, interrogation of a single particle does not suffer from the inhomogeneous broadening of the resonant line shape observed in far-field spectroscopy. This, together with the fact that binding events are monitored in a local manner, leads to an increased sensitivity, expressed via observed peak-shift with quantity of agent binding. Also, sensors based on individual particles of submicron dimensions enable at least in principle a high integration density of sensing sites for assay-like studies with high throughput. However, for this vision to come true, a suitable, parallelized addressing scheme for individual, closely spaced particles has first to be developed.

Proof-of-concept studies of single-particle sensors therefore rely on the spectroscopic determination of the plasmon resonance of an individual, sub-wavelength metallic nanoparticle. In the following, we will present four prominent optical excitation techniques suitable for this purpose - total internal reflection spectroscopy, near-field microscopy and dark-field microscopy, and photothermal imaging of very small particles with dimensions below 10 nm.

In *total internal reflection spectroscopy*, metallic nanostructures are deposited on top of a prism, and excitation takes place using illumination under total internal reflection conditions. Similar to the excitation of SPPs on a flat metal film described in chapter 3, the evanescent field above the prism acts as a local excitation source for modes at the interface, leading to resonantly enhanced scattering. This way, the frequency of spatially confined modes in

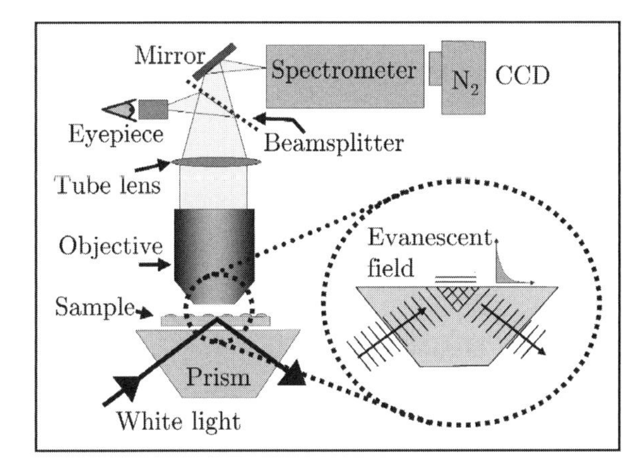

Figure 10.1. Setup for single-particle spectroscopy using evanescent excitation via total internal reflection at a prism and the monitoring of scattered light. Reprinted with permission from [Sönnichsen et al., 2000]. Copyright 2000, American Institute of Physics.

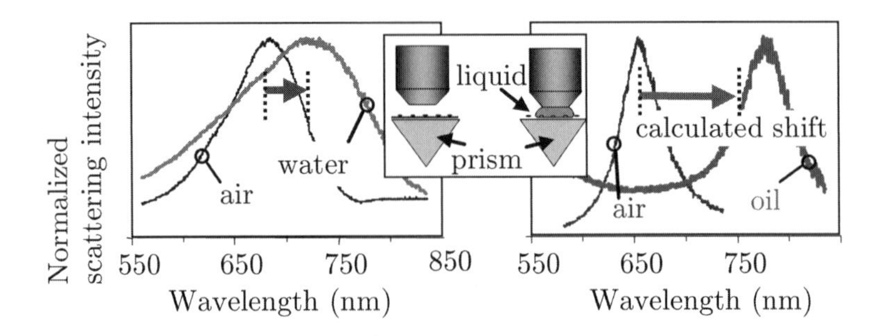

Figure 10.2. Shift of particle plasmon resonance detected using prism excitation. Reprinted with permission from [Sönnichsen et al., 2000]. Copyright 2000, American Institute of Physics.

metal nanoparticles can be determined using white light illumination and detection of the scattered light via far-field collection from the top (Fig. 10.1).

Examples of plasmon spectra of a single gold particle collected in this manner are shown in Fig. 10.2. As expected from (10.1), the resonance peak of the dipole plasmon mode red-shifts upon immersion of the particle into a high-index environment such as water or oil. The expected spectral variation of the collected intensity can in a first approximation be calculated via the formulae for the cross sections of first-order Mie theory (5.13), using the appropriate dielectric data $\varepsilon(\omega)$ for the metal. For a metal particle on a glass prism immersed in an external medium, the effective dielectric constant of the host can often simply be approximated as $1/2\left(\varepsilon_{\text{prism}} + \varepsilon_{\text{m}}\right)$.

Single-particle spectroscopy can also be performed using *near-field optical microscopy*, i.e. by placing an apertured fiber tip into the near field of the particle under study. In its simplest form, spectroscopic information is obtained by monitoring the spectral intensity distribution of radiation collected in the far field (either in transmission or reflection) ensuing from local illumination of the particle with white light. This way, the resonance frequencies and homogeneous lineshapes of plasmon modes in single particles can be determined. Pioneering spectroscopic studies of single particles have been performed using both transmission near-field optical microscopy with near-field illumination and far-field collection [Klar et al., 1998], and collection-mode near-field optical microscopy with prism-coupling illumination as described above, but with near-field instead of far-field collection [Markel et al., 1999].

In a more recent study, Mikhailovsky and co-workers have shown that transmission mode near-field optical microscopy with local white light illumination through the sub-wavelength aperture enables a high sensitivity for determining the plasmon resonance of an individual particle due to phase information encoded into the intensity collected in the far field [Mikhailovsky et al., 2004]. This is based on the fact that the light scattered by the particle in the forward direction interferes either constructively or destructively with the light directly

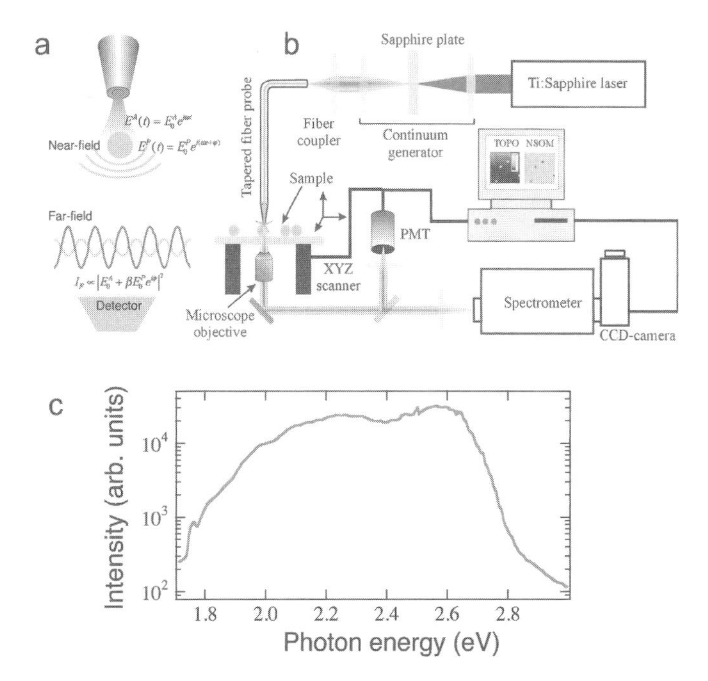

Figure 10.3. Sketch of excitation geometry (a) and experimental setup (b) for white-light illumination mode near-field optical microscopy. (c) Spectrum of the white light supercontinuum at the output of the fiber tip. Reprinted with permission from [Mikhailovsky et al., 2003]. Copyright 2003, Optical Society of America.

collected from the aperture [Batchelder and Taubenblatt, 1989]. Fig. 10.3 shows a sketch of the experimental setup and a spectra of a white light supercontinuum passing through an apertured tip. Typical examples of images of both the topography and the optical near field of gold nanoparticles are presented in Fig. 10.4a.

An investigation of the scattering and absorption process using the model of a driven damped harmonic oscillator predicts a contrast reversal of the near-field image due to the transition from destructive to constructive interference at ω_{sp} (Fig. 10.4b). We recollect from chapter 5 that, in the vicinity of the resonance, a phase shift ϕ between the driving field and the response of the electrons of π occurs, with $\phi\left(\omega_{sp}\right) = \pi/2$. An analysis of near-field images obtained at different frequencies therefore enables the determination of ω_{sp} for particles of various sizes (Fig. 10.4c).

While near-field optical extinction microscopy provides unprecedented spatial resolution for local spectroscopy, the optical probe placed in the near field of the particle poses a difficult constraint for practical sensing applications. Agent binding is additionally often monitored in a liquid environment, which

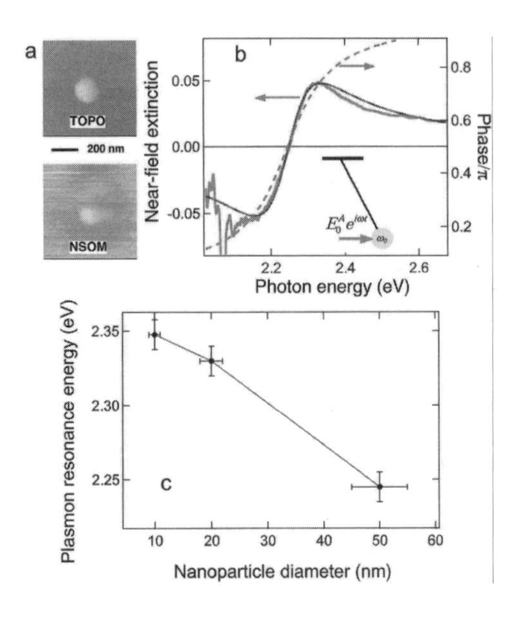

Figure 10.4. (a) Topography and near-field image of a 50 nm gold sphere. (b) Near-field extinction spectrum (solid gray curve) compared with interference (black curve) and phase (dashed curve) spectra for a single 50 nm gold particle calculated using a forced harmonic oscillator model. (c) Dependence of resonance frequency on particle size inferred from the spectra. Reprinted with permission from [Mikhailovsky et al., 2003]. Copyright 2003, Optical Society of America.

poses serious stability problems for the probe movement. Moreover, since near-field optical microscopy only allows the determination of optical properties near a surface, in situ measurements of metal nanoparticles within cell bodies are generally not possible. A more suitable geometry for such purposes is *dark-field optical microscopy*, which is a far-field technique where only light scattered by the nanoparticle is collected. Here, use of a dark-field condenser prevents the collection of the directly transmitted light. Therefore, in dark-field images metal nanoparticles appear in bright colors, defined by the resonance frequency ω_{sp} of their scattering cross section (5.13). A typical dark-field image of single gold nanoparticles is shown in Fig. 10.5c. We note that due to the constraint of the diffraction limit for focusing of the illumination spot, single-particle sensitivity can only be achieved for well-separated nanoparticles.

An example of the monitoring of a molecular binding event is shown in Figs. 10.5 and 10.6 [Raschke et al., 2003]. The coating of a gold nanoparticle with a BSA-complex leads to a slight red-shift of ω_{sp}, which is increased upon the selective binding of streptavidin molecules (Fig. 10.5b). The binding can be monitored in real-time via a recording of the resonance shift with time (Fig. 10.6), and saturation is achieved upon complete coating of the particle.

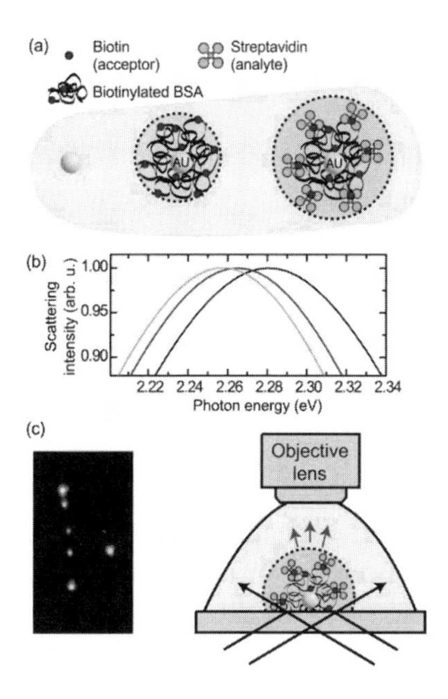

Figure 10.5. (a) Principle of a single nanoparticle biosensor monitoring the selective binding of streptavidin on an BSA-decoared gold nanoparticle. (b) Mie-theory calculations of the scattering spectra for the undecorated particle and the particle with BSA and BSA-streptavidin coating, demonstrating red-shifts of the resonance with each coating layer. (c) Dark-field pictures and sketch of the detection pathway. Reprinted with permission from [Raschke et al., 2003]. Copyright 2003, American Chemical Society.

A similar study based on single silver nanoparticles demonstrated that a sensitivity on the order of zeptomoles can be achieved, and first applications using medically-relevant assay studies are emerging [Haes et al., 2004].

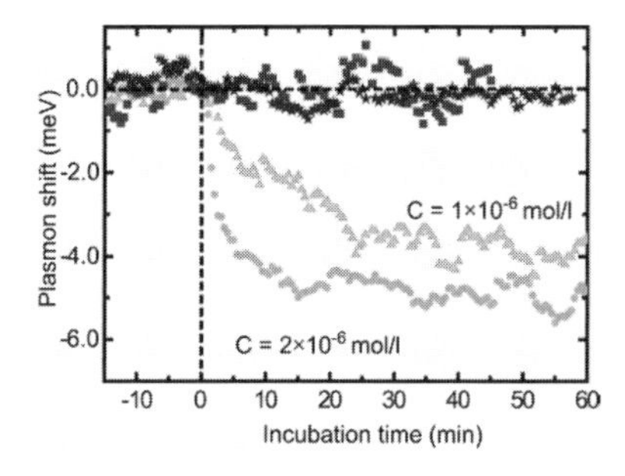

Figure 10.6. Resonance shift versus incubation time for streptavidin-BSA binding for different streptavidin concentrations C added at $t = 0$, and control experiments. Reprinted with permission from [Raschke et al., 2003]. Copyright 2003, American Chemical Society.

Further improvements in sensitivity have been predicted for single-particle sensors employing resonance line shape design, which can either be achieved using metallic nanoshells [Raschke et al., 2004], designed particle arrays with near-field coupling for hot spot generation [Enoch et al., 2004], or by using a particle-on-extended film approach to couple the particle plasmon to propagating SPPs [Chen et al., 2004]. Also, the use of elongated nanoparticles has enabled polarization-sensitive orientation sensing [Sönnichsen and Alivisatos, 2005].

The good biocompatability and well-developed surface chemistry of gold nanoparticles has further lead to their wide use in cellular imaging. In these studies, the nanoparticles mainly serve as a labeling agent for the tracking of single molecules or molecular complexes. Optical microscopy techniques such as the aforementioned dark-field illumination, differential interference contrast or total internal reflection illumination can be used for image acquisition. First in vivo studies extracting spectroscopic information akin to the particle-based studies outlined above are emerging [El-Sayed et al., 2005].

However, dark-field microscopy and other techniques relying on the detection of scattered light are not suitable for very small metal nanoparticles with diameters $d \lesssim 40$ nm immersed in a background of scatters, such as for example a biological cell. This is due to the fact that the scattering cross section decreases as d^6 with particle diameter as discussed in chapter 5. Thus, the scattering signal of particles in this small size regime is usually completely overwhelmed by larger scatterers. In order to optically pick out the signature of particles of these small sizes, a different microscopy method relying on *absorption* instead of scattering is required. Since according to Mie theory the absorption cross section scales with size only as d^3, sub-10 nm particles can be picked out of a background of bigger particles using a *photothermal* imaging technique [Boyer et al., 2002]. Fig. 10.7 shows the optical setup used in this imaging technique, consisting of a heating beam and a second, weaker probe beam detecting the absorption-induced thermal changes around the metal nanoparticles. The red probe beam is split in two parts of orthogonal polarization, and both beams are subsequently focused onto the sample to diffraction-limited spots spaced at a distance on the order of 1 μm from each other. The heating beam only overlaps with one of the probe beams, resulting in a heat-induced change in its polarization. Recombination of the two probe beams therefore leads to an intensity modulation, and via a scanning system an image of the sample under study can be constructed. In-vivo images acquired using this technique are shown in Fig. 10.8, and compared with scattering and fluorescence images for biological cells with incorporated gold nanoparticles, demonstrating the improved spatial resolution due to the detection of single particles.

Before moving on, we want to briefly mention another promising technique for the spectroscopic investigation of localized surface plasmons, based on ex-

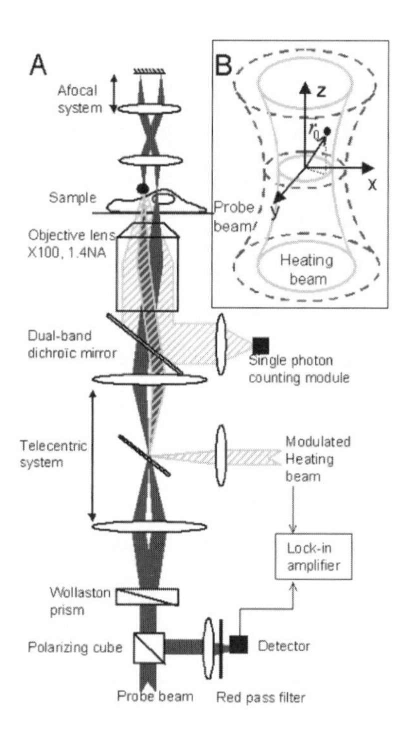

Figure 10.7. Experimental setup for photothermal imaging of very small nanoparticles. For a description see main text. Reprinted with permission from [Cognet et al., 2003]. Copyright 2003, National Academy of Sciences, U.S.A.

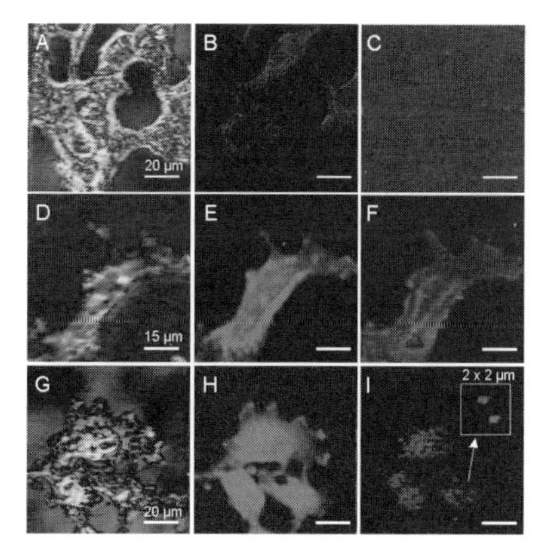

Figure 10.8. Scattering (A, D, G), fluorescence (B, E, H) and photothermal images (C, F, I) of cells. All cells are transfected with gold nanoparticles functionalized to a membrane protein (A-F in concentration 10 μg/l and for G-I the concentration is 0.5 μg/l). A-C show cells not expressing this protein, and D-I cells expressing it and thus binding the particles. The resolution is highest in pictures F and I obtained with photothermal imaging. Reprinted with permission from [Cognet et al., 2003]. Copyright 2003, National Academy of Sciences, U.S.A.

citation using electron impact. In *cathodoluminescence*, photon emission of the metal nanostructure under investigation is induced via a high-energy electron beam, and collected using a suitable detection pathway [Yamamoto et al., 2001]. As an example, Fig. 10.9 (upper part) shows the spectrum of a 140 nm silver particle excited via grazing-incidence of a 200 keV electron beam, together with a comparison with theory. Due to the large size of the particle, the signatures of both a quadrupolar and a dipolar mode are discernible. A nice feature of this technique is that by scanning of the electron beam over the particle surface, the spatial profile of the modes can be mapped out via light collection at the respective peak wavelength (Fig. 10.9, lower part). The same technique can also be used for the excitation and investigation of propagating SPPs.

All the aforementioned single-particle spectroscopy techniques are based on microscopy and thus generally not suitable for field-based sensing, e.g. in a context of environmental monitoring. Sensors based on localized particle plasmon spectroscopy amenable for such applications have been developed in the context of optical-fiber-based sensing. In a typical geometry, metal nanoparticles are spatially fixed at the end facet of an optical fiber, and the reflected light

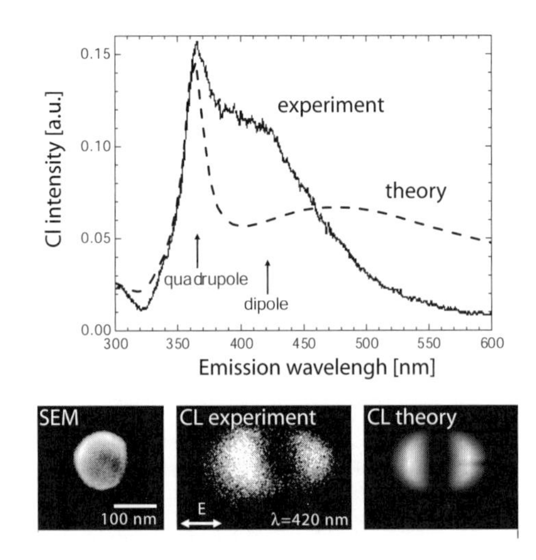

Figure 10.9. Cathodoluminescence imaging and spectroscopy of localized surface plasmons. Upper part: cathodoluminescence (CL) from a 140 nm silver particle induced by the passage of 200 keV electrons in a grazing trajectory (barely touching the particle surface). Dipolar and quadrupolar components can be separated in the spectrum. Lower part, from left to right: SEM image of the particle under consideration; CL rate as a function of the position of the electron beam, which is scanned over the particle, for an emission wavelength corresponding to the dipolar feature of the spectrum; theoretical prediction for the latter. Figure courtesy of N. Yamamoto and F. J. García de Abajo, personal communication.

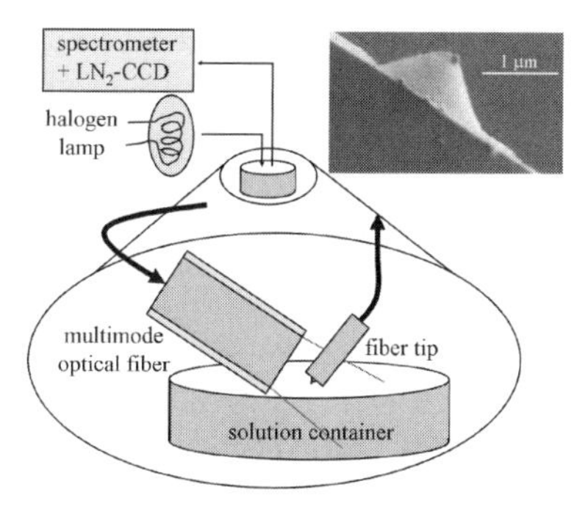

Figure 10.10. Optical setup for measuring the scattering of a single nanoparticle in various solvents through an optical fiber. The inset shows a SEM image of a nanoparticle attached to the fiber tip. Reprinted with permission from [Eah et al., 2005]. Copyright 2005, American Institute of Physics.

upon white-light illumination collected through the fiber in reflection using an inperfect splice and spectrally resolved [Mitsui et al., 2004]. Immersion of the particle-decorated end facet into the environment under study then allows refractive index sensing of gaseous or liquid agents.

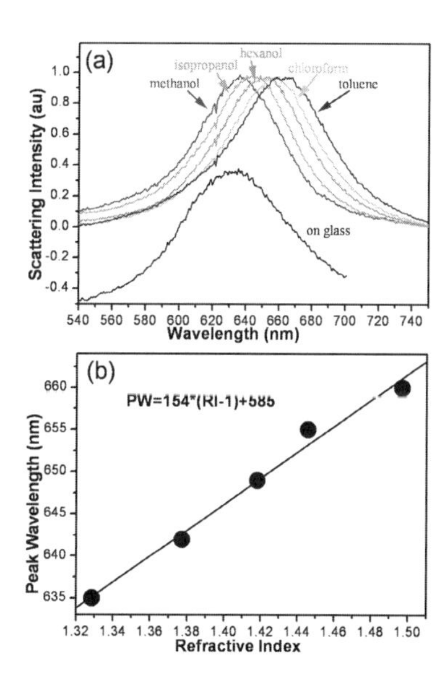

Figure 10.11. (a) Normalized scattering spectra of a single gold nanoparticle in various solvents measured through the fiber. (b) Dependence of the resonance position on the index of refraction of the solvent. Reprinted with permission from [Eah et al., 2005]. Copyright 2005, American Institute of Physics.

Eah and co-workers have recently demonstrated single-particle sensitivity using this technique [Eah et al., 2005]. Fig. 10.10 shown a schematic of the optical setup. A single gold nanoparticle is fixed at the end facet of a sharp fiber tip via direct pick-up from a flat surface covered with metal colloids. In this study, external illumination was used via a second multimode fiber, and the scattered signal collected via the fiber tip. Typical spectra for immersion in a variety of solutions of different refractive index are shown in Fig. 10.11.

10.2 Surface-Plasmon-Polariton-Based Sensors

The vast majority of surface plasmon sensing work carried out so far has not been based on the spectroscopic determination of the particle plasmon resonance, but on the interrogation of propagating SPP waves at a metal/air interface. Using surface functionalization, agent-specific binding can be achieved, changing the refractive index of the metal surface superstrate and thus the dispersion relation of the propagating SPPs. Binding events can then be monitored by studying the changing phase-matching condition via either wavelength- or angular interrogation. Historically, for sensing applications both prism-coupling and grating-coupling techniques as described in chapter 3 have been preferred for SPP excitation via light beams. A review of these techniques in a sensing context was recently conducted by Homola and colleagues [Homola et al., 1999].

Since both grating and prism coupling have been extensively discussed in chapter 3, and due to the simplicity of their employment in sensing appliations, we only want to comment here on a few extensions of these standard techniques, with particular promise in terms of enhancement of sensing sensitivity. In general, the performance of a SPP-based sensor increases with the amount of field confinement, and also the magnitude of the attenuation length L (note that an increase in one leads to a decrease in the other). As an example of the use of structures with low SPP attenuation, multilayer geometries have proved highly useful for sensing purposes, and enhanced sensitivity using long-ranging modes excited via prism coupling geometries have been reported [Nenninger et al., 2001].

A further improvement in sensitivity can be achieved by exploiting the fact that in the prism coupling geometry, the phase of the reflected field changes as the phase-matching condition for SPP excitation is transversed, in analogy to the discussion of phase-sensitive near-field imaging of localized modes in the previous section. Using an input consisting of both TE and TM beam components, Hooper and Sambles demonstrated a highly sensitive device capable of measuring refractive index changes of 2×10^{-7} [Hooper and Sambles, 2004b]. The experimental setup, based on polarization dithering of the input beam to enable a differential detection of changes to the polarization ellipse, is shown in Fig. 10.12. In this case, phase changes of the TM-polarized component of the

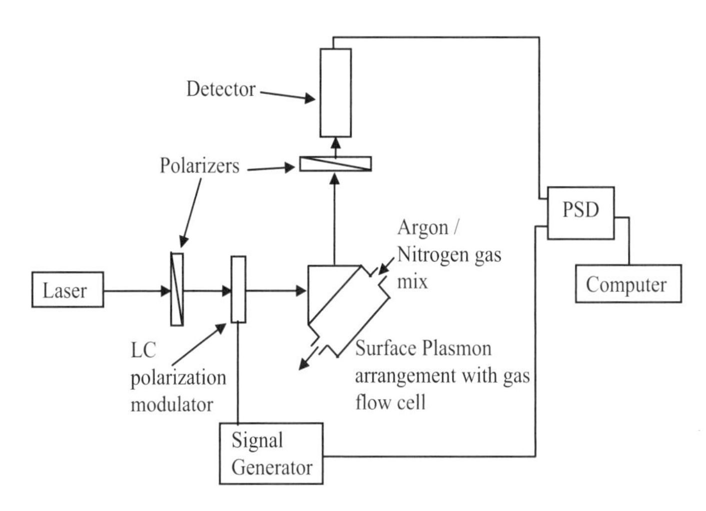

Figure 10.12. Experimental setup for differential ellipsometric detection of refractive index changes using SPPs on a metal film excited via prism coupling. Reprinted with permission [Hooper and Sambles, 2004b]. Copyright 2004, American Institute of Physics.

input beam induced by changes in the refractive index of the superstrate manifest themselves via polarization changes in the reflected light beam. Fig. 10.13 shows results on the obtained polarization rotation, depending upon the ratio of two gases in a mixture.

While SPP excitation using prism or grating coupling is a convenient method of choice for proof-of-concept demonstrations of SPP sensors, *waveguide SPP sensors* employing phase-matching between a waveguide mode in a guiding

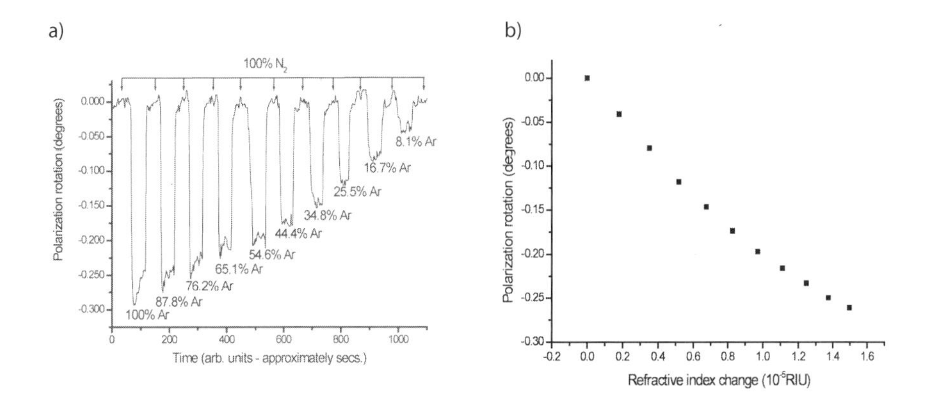

Figure 10.13. (a) Polarization rotation for varying gas ratios. (b) Polarization rotation as a function of refractive index. Reprinted with permission from [Hooper and Sambles, 2004b]. Copyright 2004, American Institute of Physics.

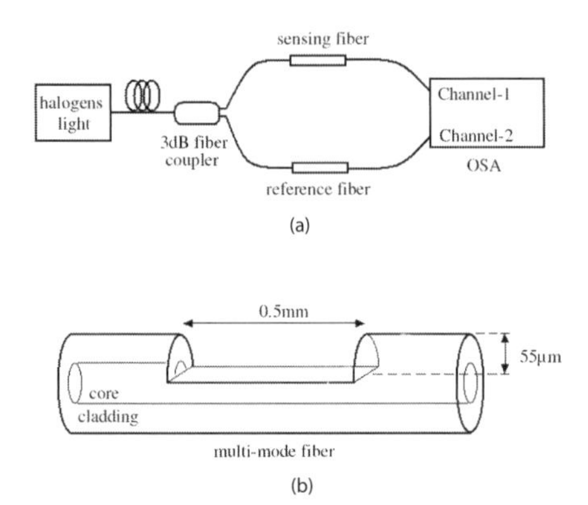

Figure 10.14. SPP sensor based on a multimode optical fiber. (a) Sketch of the sensing system consisting of side-polished sensing and reference fibers. (b) Sketch of the side-polished fibers. Reprinted with permission from [Tsai et al., 2005]. Copyright 2005, Optical Society of America.

layer beneath the exposed metal surface are favorable from an integration point of view. A particularly interesting device with possibilities for field-use is the optical fiber SPP resonance sensor [Slavik et al., 1999]. In its usual form, such a sensor consists of a (single- or multimode) optical fiber, one side of which has been polished away to expose the core. The coating of this region with a thin metal layer then allows the excitation of SPPs via the core-guided mode(s), and their signature can be detected by monitoring the light guided past the interaction region [Homola et al., 1997]. The simplicity of this approach has made fiber excitation the method of choice for many SPP sensing studies.

A sketch of a typical sensing region cut into a multimode optical fiber is shown in Fig. 10.14b. Exposure of the core can be accomplished via the aforementioned polishing, or etching and also tapering techniques. Using a white light illumination source and thus wavelength selectivity is a particularly appealing approach, since modern sources such as fiber-based supercontinuum sources allow for easy integration directly into the sensing fiber. In order to improve the sensitivity, a combination of a reference and sensing fiber (Fig. 10.14a) can be employed to enable either interferometric detection or difference-signal analysis [Tsai et al., 2005].

In the example presented here, both the sensing and the reference fiber are side-polished and metalized with a 40 nm gold layer. The reference fiber is immersed into distilled water, and the sensing fiber in a liquid of different refractive index. SPP spectra from both arms are recorded, and the difference in light intensity versus wavelength determined (Fig. 10.15a). A difference of zero corresponds to the crossing point between the two SPP curves, which exhibits a strong dependence on the refractive index difference, as shown in

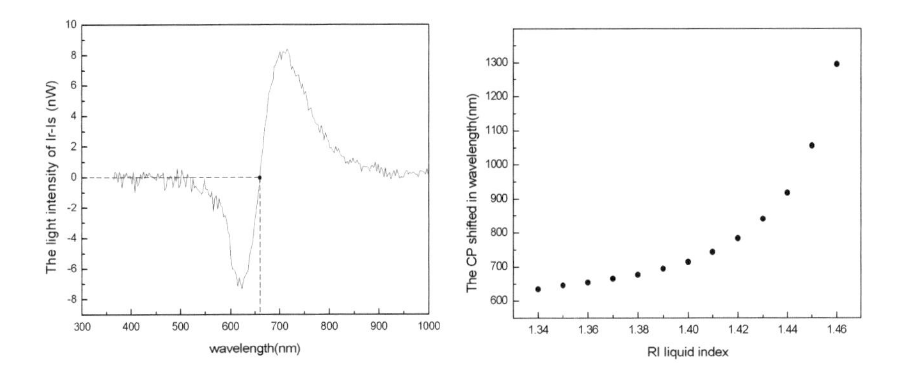

Figure 10.15. (a) Difference in light intensity between the sensing and the reference arm vs. optical wavelength using the SPP fiber sensor structure of Fig. 10.14. Here, the sensing arm is immersed in alcohol, and the reference fiber in distilled water. (b) Experimental results for the shift in crossing-point wavelength of the two SPP spectra versus refractive index. Reprinted with permission from [Tsai et al., 2005]. Copyright 2005, Optical Society of America.

Fig. 10.15b. A high sensitivity for refractive index sensing on the order of 10^{-6} can be achieved. Improved designs of the geometry of the sensing region enabled by advancements in polishing [Zhang et al., 2005] and tapering techniques [Kim et al., 2005] continuously push the obtained sensitivity limits, placing SPP sensors at the forefront of optical sensing techniques.

SPPs can also be excited using optical fibers coated homogeneously with a concentric metal layer. For thin tapers, this lead to the generation of hybrid fiber-SPP modes with interesting properties [Al-Bader and Imtaar, 1993, Prade and Vinet, 1994]. We cannot go into the details of these hybrid modes here, but want to point out that they have indeed been recently observed [Diez et al., 1999], and that applications as sensors have been demonstrated [Monzon-Hernandez et al., 2004].

Chapter 11

METAMATERIALS AND IMAGING WITH SURFACE PLASMON POLARITONS

The notion that the electromagnetic response of a material can be engineered via periodic variations in structure and composition has been extensively investigated over the last two decades. A well-known example are *photonic crystals*, dielectric materials with a periodic modulation of their (real) refractive index $n = \sqrt{\varepsilon}$, achieved via the inclusion of scattering elements such as holes of different dielectric constant into the embedding host. This way, the dispersion relation for electromagnetic waves propagating through the artificial crystal can be engineered, and band gaps in frequency space established that inhibit propagation. In photonic crystals, both the size and the periodicity of the index modulations are of the order of the wavelength λ in the material. We have seen in chapter 7 that the SPP analogue of this concept, a metal surface with a periodic lattice of surface protrusions, enables control over SPP propagation.

An equally intriguing possibility for designing artificial materials with a controlled photonic response are *metamaterials*. In contrast to photonic crystals, in this case both the size and the periodicity of the scattering elements are significantly smaller than λ. Therefore, they can in a sense be viewed as microscopic building blocks of an artificial material, in analogy to atoms in conventional materials found in nature. Using the same reasoning applied to the transitioning from the microscopic to the macroscopic form of Maxwell's equations, the electromagnetic response of a metamaterial can be described via both an *effective* permittivity $\varepsilon(\omega)$ and permeability $\mu(\omega)$. Since on the sub-wavelength scale the electric and the magnetic fields are essentially decoupled, $\varepsilon(\omega)$ and $\mu(\omega)$ can often be controlled independently by the use of appropriately shaped scatterers.

The corrugated perfectly-conducting surfaces described in chapter 6 are an example of a metamaterial with an engineered electric response $\varepsilon(\omega)$. We have seen that such an interface can be described as an effective medium, with a plasma frequency ω_p controlled by the geometry. In the first part of the current chapter, we will briefly describe other prominent examples of metamaterials, specifically focusing on how a magnetic response can be achieved using sub-wavelength arrangements of non-magnetic constituents. Appropriate materials design allows both $\varepsilon(\omega)$ and $\mu(\omega)$ to be negative in a certain frequency range, leading to a *negative refractive index* $n = \sqrt{\mu\varepsilon}$[1].

The rich physics of metamaterials and specifically those with a negative refractive index will only be briefly discussed, with a view to the challenges of creating $n < 0$ at optical frequencies. We will see that arrangements of metal nanoparticles sustaining localized plasmon resonances are a promising route for creating such structures. For a more detailed exploration of metamaterials, we refer the reader to specialized reviews such as [Smith et al., 2004] as a starting point.

One of the most intriguing possibilities of negative index materials is imaging with sub-wavelength resolution, which has become known under the paradigm of the *perfect lens*. The second part of this chapter addresses efforts to demonstrate this effect at optical frequencies via the use of SPP excitations in thin metal films.

11.1 Metamaterials and Negative Index at Optical Frequencies

The metamaterial concept of creating composites with desired electromagnetic properties has already enabled new possibilities for the control of electromagnetic radiation in the THz and microwave region of the spectrum. We have discussed in chapter 6 in detail how appropriate sub-wavelength structuring of a metal surface can lead to a geometry-defined plasma frequency ω_p in this frequency region. Another prominent example of a metamaterial sustaining low-frequency plasmons is a regular three-dimensional lattice of metal wires with micron-size diameter [Pendry et al., 1996]. It can be shown that the electric response of such a structure can be viewed as that of an effective medium with a free electron density determined by the fraction of space occupied by the wires. As with the structures described in chapter 6, the effective $\varepsilon(\omega)$ of the wire lattice is of the plasma form (1.20), with ω_p lowered into the microwave range for an appropriate mesh size. The dielectric response of the wire lattice to microwave radiation is similar to that of a metal at optical frequencies.

One motivation of metamaterials design is therefore to shift electric resonances of natural materials (particularly metals), expressed via $\varepsilon(\omega)$, to lower frequencies. The other motivation is in the opposite direction: The creation

[1] It can be shown that the negative sign of the square root has to be chosen, since in such a material the phase and group velocities of the transmitted radiation point in opposite directions.

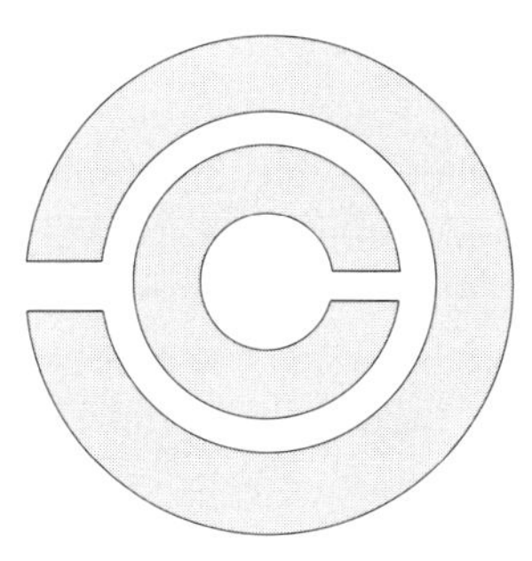

Figure 11.1. Sketch of a split ring resonator for engineering the magnetic permeability $\mu(\omega)$ of a metamaterial.

of magnetic resonances, described by $\mu(\omega)$, at frequencies higher than those present in naturally-occurring magnetic materials. More specifically, the region of interest lies between the THz and the visible parts of the spectrum.

Whereas the magnetism of inherently magnetic materials is caused by unpaired electron spins [Kittel, 1996], the magnetism of metamaterials is entirely due to geometry-induced resonances or plasmonic effects of their sub-wavelength building blocks. A particularly useful geometry is that of the split ring resonator, depicted in Fig. 11.1 in its most simple form. It consists of two planar concentric conductive rings, each with a gap. Pendry and co-workers have shown that a regular array of these structures, with both structure size and lattice constant of dimensions much smaller than the wavelength region of interest, can exhibit a magnetic response [Pendry et al., 1999].

In a simplified view, a time-varying magnetic field induces a magnetic moment in a split ring resonator via the induction of currents flowing in circular paths. This inherently weak response is magnified via a resonance: the structure acts as a sub-wavelength LC circuit with inductance L and capacitance C. Therefore, the magnetic permeability μ exhibits a resonance at $\omega_{\mathrm{LC}} = 1/\sqrt{LC}$. Intriguingly, as is typical for a resonant process, for frequencies right above ω_{LC}, $\mu < 0$. As will be discussed below, combined with wire arrays this allows the creation of metamaterials exhibiting both negative permittivity and permeability, and thus a negative refractive index as described in the introduction.

Following initial demonstrations for microwave frequencies (reviewed in [Smith et al., 2004]), metamaterials with a magnetic response engineered using split ring resonators were demonstrated in the THz regime by Yen and co-workers [Yen et al., 2004]. The effective permeability of the metamaterial determined from measurements can be described using a Lorentz term

$$\mu(\omega) = 1 - \frac{F\omega^2}{\omega^2 - \omega_{\mathrm{LC}}^2 + i\Gamma\omega}, \tag{11.1}$$

where ω_{LC} is the resonance frequency and F a geometrical factor. Γ describes resistive losses in the split ring resonator. As for a typical resonance process, for $\omega \ll \omega_{LC}$ the induced magnetic dipole is in phase with the excitation field. In this region, the metamaterial therefore exhibits a paramagnetic response. For increasing frequencies, the currents start to lag behind the driving field, and for $\omega \gg \omega_{LC}$ the dipole response is completely out of phase with the driving field. In this region, the metamaterial is diamagnetic ($\mu < 1$). For the frequency region just above ω_{LC}, the permeability is negative ($\mu < 0$). We note that the magnetic dipole is an *induced dipole* only - no permanent magnetic moment is present.

This discussion of metamaterials with an engineered electric or magnetic response suggests that a material consisting of a lattice of both split ring resonators and metal wires or rods should exhibit a frequency region where both $\varepsilon < 0$ and $\mu < 0$, implying $n < 0$. Shelby and co-workers demonstrated such a negative-index metamaterial at microwave frequencies [Shelby et al., 2001]. Using a wedge-shaped structure, negative refraction (a consequence of a negative refractive index) was confirmed [Smith et al., 2004]. While the metamaterial used in this study was of a three-dimensional nature, inherently planar structures consisting of split ring resonators and rods working at THz frequencies have been successfully fabricated using microfabrication techniques [Moser et al., 2005].

For microwave and THz frequencies, metamaterials such as the ones described above consisting of conductive materials show a simple size scaling of their resonance frequencies, i.e. $\omega_{LC} \propto 1/a$, where a is the typical size of a split ring resonator. However, this scaling breaks down for higher frequencies, where the response of the metal becomes less and less ideal, and the kinetic energy of the electrons needs to be taken into account. Theoretical investigations have suggested that this leads to a saturation of the increase of ω_{LC} with frequency for $f > 100$ THz ($\lambda_0 < 3$ μm) [Zhou et al., 2005]. Using gold split ring resonators of a minimum feature size of 35 nm, Klein and co-workers have shown that the resonance in μ can be pushed down to a wavelength $\lambda = 900$ nm in the near-infrared. It is at this point not clear how much the resonance frequency can be increased into the visible regime using this concept.

Apart from split ring resonators, rod-shaped structures can also be used to create a material with negative refractive index in the near-infrared. Shalaev and co-workers demonstrated $n = -0.3$ at $\lambda = 1.5$ μm using a metamaterial consisting of rod-shaped gold/insulator/gold sandwich structures [Shalaev et al., 2005]. Fig. 11.2 shows a schematic and a SEM image of the composite rod structure and the metamaterial lattice. Each rod consists of a 50 nm SiO_2 layer sandwiched between two 50 nm gold layers. As in our discussion of split ring resonators, the magnetic response can be thought to arise from a resonance in the LC circuit consisting of the bottom and top gold layer, as

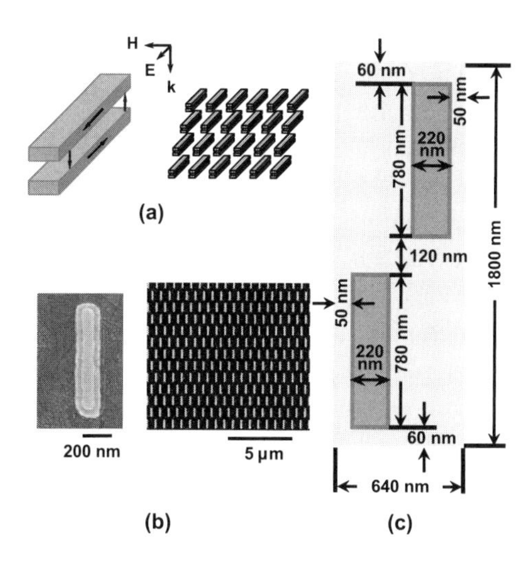

Figure 11.2. (a) Schematic and (b) SEM image of a planar metamaterial consisting of pairs of parallel gold nanorods. (c) Sketch of the unit cell of this structure. Reprinted with permission from [Shalaev et al., 2005]. Copyright 2005, Optical Society of America.

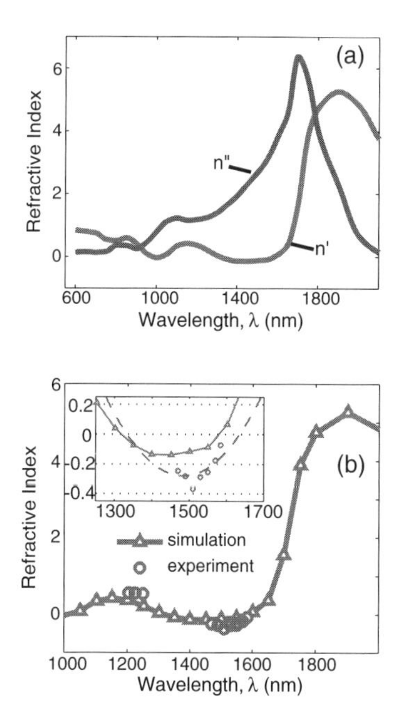

Figure 11.3. (a) Real and imaginary parts n', n'' of the refractive index for the metamaterial of Fig. 11.2 determined using simulations. (b) Comparison between simulations (triangles) and experimentally determined values (circles) of the real part of the refractive index. The inset shows a magnified view of the region of negative refractive index. Copyright with permission from [Shalaev et al., 2005]. Copyright 2005, Optical Society of America.

symbolized in Fig. 11.2a. The metal rods provide the inductance, and the insulating spacer layer the capacitance. The refractive index of this metamaterial in the near-infrared range of the spectrum is shown in Fig. 11.3. Around the telecommunication window at $\lambda = 1500$ nm, $n < 0$. We note that contrary to the metamaterials discussed so far, the dimensions of the unit cell (Fig. 11.2c) are of the order of the wavelength. Also, as with split ring resonators, a simple linear scaling with size towards higher frequencies in the visible regime should be prevented both by increasing losses and the importance of plasmonic effects.

In another study, a metamaterial with negative permeability in the visible part of the spectrum was demonstrated by Grigorenko and co-workers. In this case, the metamaterial consists of pairs of dome-shaped gold nanoparticles [Grigorenko et al., 2005]. The pairs essentially act as small bar magnets, and antisymmetric coupling of the localized plasmon resonances gives rise to cancellation of the magnetic component of the incident field, thus yielding $\mu < 0$. Also, an approach has recently been suggested based on U-shaped metal nanoparticles, making active use of the plasmonic response instead of LC-effects, which should provide $n < 0$ at optical frequencies [Sarychev et al., 2006]. Research in this field is going on at a breathtaking pace, and we can expect significant advances in the coming years.

11.2 The Perfect Lens, Imaging and Lithography

We want to finish this chapter by briefly discussing another fascinating consequence of materials with a negative index of refraction, namely the possibility of a perfect lens [Pendry, 2000, Smith et al., 2004]. In 2000, Pendry showed that a slab of an ideal (lossless) material with $n = -1$ can reproduce a perfect image of an object *placed into the near-field* on one side of the slab at an equal distance on the other side. Due to the property of negative refraction, it can easily be shown that light from a point source on one side of the

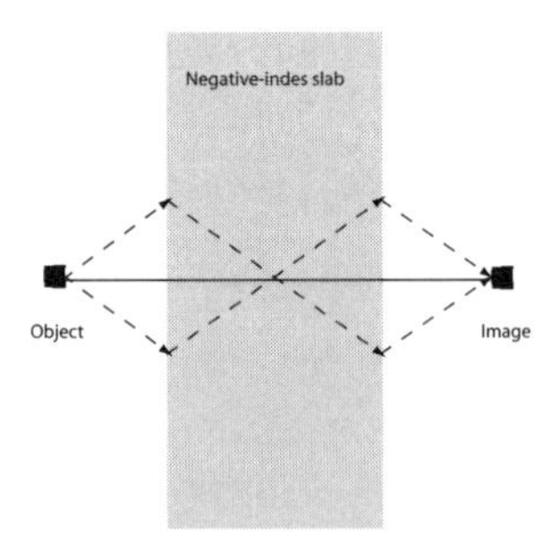

Figure 11.4. Schematic of the planar negative-index lens. Light diverging from a point source converges back towards a point in the negative-index medium due to negative refraction. On the other side of the planar slab, another focus is formed.

negative-index slab should come to a focus on the other side, as sketched in Fig. 11.4. It is more suprising however that all Fourier components of the two-dimensional object, not only those fulfilling the condition $k_x^2 + k_y^2 < \omega^2/c^2$, can be reproduced in the image plane. This is due to resonant amplification of the evanescently decaying components of the image via surface modes of the negative-index slab [Pendry, 2000].

At optical frequencies and for a slab thickness $d \ll \lambda$, the electrostatic limit applies and the electric and magnetic fields are essentially decoupled. It can be shown that in this case the requirement $\mu < 0$ can be relaxed, and sub-resolution imaging thus achieved using materials where only $\text{Re}\,[\varepsilon] < 0$, namely metals. Note however that due to attenuation ($\text{Im}\,[\varepsilon] > 0$), some of the high-resolution information is expected to get lost during the imaging process, and the image will thus not be perfect anymore. It was suggested that this *poor man's lens* could be achieved with a thin film of silver.

Here, the evanescent components of the object fields are resonantly amplified via coupling to SPPs sustained by the silver film. An experimental setup for demonstrating sub-wavelength imaging is shown in Fig. 11.5. In this study, an image etched into a chrome mask is transferred onto a photoresist via a thin silver layer [Fang et al., 2005]. Images demonstrating the achieved resolution and results from a control experiment where the silver film was replaced by a polymer layer are presented in Fig. 11.6. While the 40 nm width of the object letters was not reproduced, a significant increase in resolution in the presence of the silver layer is apparent. Additional studies of both single [Melville and

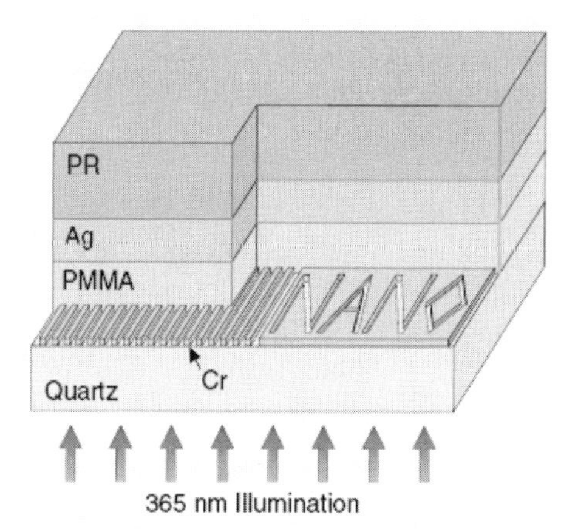

Figure 11.5. Schematic of an optical superlens. A 35 nm thin silver imaging layer is separated from a chrome mask via a 40 nm polymer layer. Upon illumination of the chrome mask with UV light, an image mediated by the silver layer is recorded in a thin photoresist film. Reprinted with permission from [Fang et al., 2005]. Copyright 2005, AAAS.

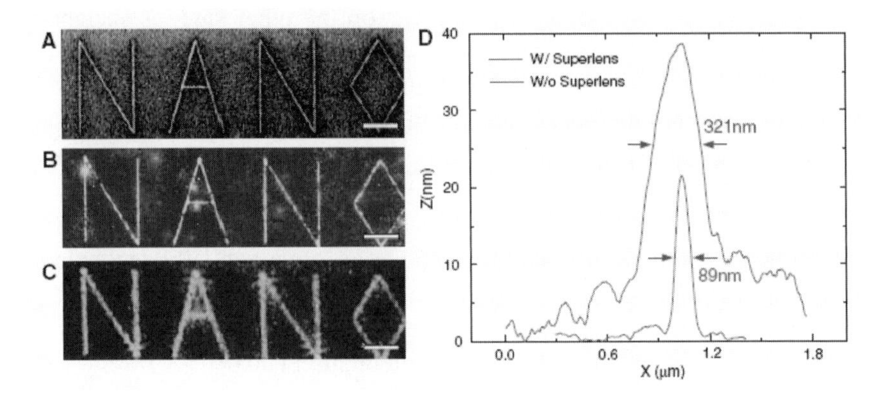

Figure 11.6. (a) FIB image of the object plane. The linewidth of the letters is about 40 nm. (b,c) AFM image of the developed photoresist with the silver imaging layer (Fig. 11.5) present (b) or replaced (c) by a PMMA layer. (d) Averaged cross section of the letter "A" with and without the lens. Reprinted with permission from [Fang et al., 2005]. Copyright 2005, AAAS.

Blaikie, 2005] and double-layer [Melville and Blaikie, 2006] silver structures have confirmed the resolution-enhancing properties.

It is anticipated that this concept could be of use for applications in lithography, where direct contact between the photoresist layer and the object mask is often undesirable. However, given the resolution constraints imposed by the conductive losses in the imaging layer, it is an open question whether this will be viable in a technological context. As a last note, we want to point out that also the design of masks with features sustaining localized plasmon resonances [Srituravanich et al., 2004, Luo and Ishihara, 2004] has been suggested for lithography beyond the classical resolution limit imposed by diffraction. In this case, the enhanced near field due to the localized modes leads to enhanced exposure of an adjacent resist layer.

Chapter 12

CONCLUDING REMARKS

Plasmonics is a fascinating and currently vastly expanding area of research, and hopefully reading through this text has provided the interested reader not only with an overview, but also with a solid foundation for own explorations. Clearly, the diversity of emerging and potential applications of sub-wavelength optics with metals together with successful proof-of-concept studies suggest that interest in the field will be soaring for many years to come.

So where to go from here? For virtually all aspects of plasmonics described in this book, specialized review articles exist within the scientific literature. Especially for areas such as sensing or metamaterials that could only be described without going into a great amount of detail, the excellent reviews available should be consulted. Apart from that, original publications such as the ones cited in the reference section are an invaluable resource for further literature studies.

I very much hope that this book will serve its purpose to educate and attract people to this fascinating area of nanophotonics. Any suggestions for improvements of this text are most welcome.

References

Adam, P. M., Salomon, L., de Fornel, F., and Goudonnet, J. P. (1993). Determination of the spatial extension of the surface-plasmon evanescent field of a silver film with a photon scanning tunneling microscope. *Phys. Rev. B*, 48(4):2680–2683.

Al-Bader, S. J. and Imtaar, M. (1993). Optical fiber hybrid-surface plasmon polaritons. *J. Opt. Soc. Am. B*, 10(1):83–88.

Andreani, L. C., Panzarini, G., and Gérard, J.-M. (1999). Strong-coupling regime for quantum boxes in pillar microcavities: Theory. *Phys. Rev. B*, 60(19):13276.

Anger, Pascal, Bharadwaj, Palash, and Novotny, Lukas (2006). Enhancement and quenching of single-molecule fluorescence. *Phys. Rev. Lett.*, 96:113002.

Antoine, Rodolphe, Brevet, Pierre F., Girault, Hubert H., Bethell, Donald, and Schiffrin, David J. (1997). Surface plasmon enhanced non-linear optical response of gold nanoparticles at the air/toluene interface. *Chem. Commun.*, pages 1901–1902.

Ashcroft, Neil W. and Mermin, N. David (1976). *Solid state physics*. Saunders College Publishing, Orlando, FL, first edition.

Avrutsky, Ivan (2004). Surface plasmons at nanoscale relief gratings between a metal and a dielectric medium with optical gain. *Phys. Rev. B*, 70:155416.

Babadjanyan, A. J., Margaryan, N. L., and Nerkarayan, Kh. V. (2000). Superfocusing of surface polaritons in the conical structure. *J. Appl. Phys.*, 87(8):3785–3788.

Bai, Benfeng, Li, Lifeng, and Zeng, Lijiang (2005). Experimental verification of enhanced transmission through two-dimensionally corrugated metallic films without holes. *Opt. Lett.*, 30(18):2360–2362.

Baida, F. I. and van Labeke, D. (2002). Light transmission by subwavelength annular aperture arrays in metallic films. *Opt. Commun.*, 209:17–22.

Bakker, Reuben M., Drachev, Vladimir P., Yuan, Hsiao-Kuan, and Shalaev, Vladimir M. (2004). Enhance transmission in near-field imaging of layered plasmonic structures. *Opt. Express*, 12(16):3701–3706.

Barnes, W. L. (1999). Electromagnetic Crystals for Surface Plasmon Polaritons and the Extraction of Light from Emissive Devices. *J. Lightwave Tech.*, 17(11):2170–2182.

Barnes, W. L., Murray, W. A., Dintinger, J., Devaux, E., and Ebbesen, T. W. (2004). Surface plasmon polaritons and their role in the enhanced transmission of light through periodic arrays of subwavelength holes in a metal film. *Phys. Rev. Lett.*, 92(10):107401.

Batchelder, J. S. and Taubenblatt, M. A. (1989). Interferometric detection of forward scattered light from small particles. *Appl. Phys. Lett.*, 55(3):215–217.

Baumberg, Jeremy J., Kelf, Timothy A., Sugawara, Yoshihiro, Cintra, Suzanne, Abdelsalam, Mamdouh E., Bartlett, Phillip N., and Russell, Andrea E. (2005). Angle-resolved surface-enhanced Raman scattering on metallic nanostructured plasmonic crystals" . *Nano Letters*, 5(11):2262–2267.

Berini, P. (1999). Plasmon-polariton modes guided by a metal film of finite width. *Opt. Lett.*, 24(15):1011–1013.

Berini, P. (2000). Plasmon-polariton waves guided by thin lossy metal films of finite width: Bound modes of symmetric structures. *Phys. Rev. B*, 61(15):10484.

Berini, P. (2001). Plasmon-polariton waves guided by thin lossy metal films of finite width: Bound modes of asymmetric structures. *Phys. Rev. B*, 63(12):125417.

Bethe, H. A. (1944). Theory of diffracion by small holes. *Phys. Rev.*, 66(7–8):163–182.

Bloembergen, N., Chang, R. K., Jha, S. S., and Lee, C. H. (1968). Optical second-harmonic generation in reflection from media with inversion symmetry. *Phys. Rev.*, 174(3):813–822.

Bohren, Craig F. and Huffman, Donald R. (1983). *Absorption and scattering of light by small particles*. John Wiley & Sons, Inc., New York, NY, first edition.

Bonod, Nicolas, Enoch, Stefan, Li, Lifeng, Popov, Evgeny, and Nevière, Michel (2003). Resonant optical transmission through thin metallic films with and without holes. *Opt. Express*, 11(5):482–490.

Borisov, A. G., de Abajo, F. J. García, and Shabanov, S. V. (2005). Role of electromagnetic trapped modes in extraordinary transmission in nanostructured metals. *Phys. Rev. B*, 71:075408.

Bouhelier, A., Huser, Th., Tamaru, H., Güntherodt, H.-J., Pohl, D. W., Baida, Fadi I., and Labeke, D. Van (2001). Plasmon optics of structured silver films. *Phys. Rev. B*, 63:155404.

Bouhelier, A. and Wiederrecht, G. P. (2005). Surface plasmon rainbow jets. *Opt. Lett.*, 30(8):884–886.

Bouwkamp, C. J. (1950a). On Bethe's theory of diffraction by small holes. *Philips Research Reports*, 5(5):321–332.

Bouwkamp, C. J. (1950b). On the diffraction of electromagnetic waves by small circular disks and holes. *Philips Research Reports*, 5(6):401–422.

Bouwkamp, C. J. (1954). Diffraction theory. *Rep. Prog. Phys.*, 17:35–100.

Boyd, Robert W. (2003). *Nonlinear Optics*. Academic Press, San Diego, CA, second edition.

Boyer, D., Tamarat, P., Maali, A., Lounis, B., and Orrit, M. (2002). Photothermal imaging of nanometer-sized metal particles among scatterers. *Science*, 297:1160–1163.

Bozhevolnyi, S. I., , Volkov, Valentyn S., Devaux, Eloise, Laluet, Jean-Yves, and Ebbesen, Thomas W. (2006). Channel plasmon subwavelength waveguide components including interferometers and ring resonators. *Nature*, 440:508–511.

Bozhevolnyi, S. I., Beermann, Jonas, and Coello, Victor (2003). Direct observation of localized second-harmonic enhancement in random metal nanostructures. *Phys. Rev. Lett.*, 90(19):197403.

Bozhevolnyi, S. I., Erland, J., Leosson, K., Skovgaard, P. M. W., and Hvam, J. M. (2001). Waveguiding in surface plasmon polariton band gap structures. *Phys. Rev. Lett.*, 86:3008–3011.

Bozhevolnyi, Sergey I., Nikolajsen, Thomas, and Leosson, Kristjan (2005a). Integrated power monitor for long-range surface plasmon waveguides. *Opt. Commun.*, 255:51–56.

Bozhevolnyi, Sergey I., Volkov, Valentyn S., Devaux, Eloise, and Ebbesen, Thomas W. (2005b). Channel plasmon-polariton guiding by subwavelength metal grooves. *Phys. Rev. Lett.*, 95:046802.

Bravo-Abad, J., García-Vidal, F. J., and Martín-Moreno, L. (2004a). Resonant transmission of light through finite chains of subwavelength holes in a metallic film. *Phys. Rev. Lett.*, 93:227401.

Bravo-Abad, J., Martín-Moreno, L., and García-Vidal, F. J. (2004b). Transmission properties of a single metallic slit: From the subwavelength regime to the geometrical-optics limit. *Phys. Rev. E*, 69:026601.

Brongersma, Mark L., Hartman, John W., and Atwater, Harry A. (2000). Electromagnetic energy transfer and switching in nanoparticle chain arrays below the diffraction limit. *Phys. Rev. B*, 62(24):R16356–R16359.

Burke, J. J. and Stegeman, G. I. (1986). Surface-polariton-like waves guided by thin, lossy metal films. *Phys. Rev. B*, 33(8):5186–5201.

Cao, Hua, Agrawal, Amit, and Nahata, Ajay (2005). Controlling the transmission resonance lineshape of a single subwavelength aperture. *Opt. Express*, 13(3):763–769.

Chan, H. B., Marcet, Z., Woo, Kwangje, Tanner, D. B., Carr, D. W., Bower, J. E., Cirelli, R. A., Ferry, E., Klemens, F., Miner, J., Pai, C. S., and Taylor, J. A. (2006). Optical transmission through double-layer metallic subwavelength slit arrays. *Opt. Lett.*, 31(4):516–518.

Chang, Shih-Hui, Gray, Stephen K., and Schatz, George C. (2005). Surface plasmon generation and light transmission by isolated nanoholes and arrays of nanoholes in thin metal films. *Opt. Express*, 13(8):3150–3165.

Charbonneau, Robert, Berini, Pierre, Berolo, Ezio, and Lisicka-Shrzek, Ewa (2000). Experimental observation of plasmon-polariton waves supported by a thin metal film of finite width. *Opt. Lett.*, 25(11):844.

Charbonneau, Robert, Lahoud, Nancy, Mattiussi, Greg, and Berini, Pierre (2005). Demonstration of integrated optics elements based on long-ranging surface plasmon polaritons. *Opt. Express*, 13(3):977–983.

Chau, K. J., Dice, G. D., and Elezzabi, A. Y. (2005). Coherent plasmonic enhanced terahertz transmission through random metallic media. *Phys. Rev. Lett.*, 94:173904.

Chen, C. K., Heinz, T. F., Ricard, D., and Shen, Y. R. (1983). C. K. Chen and T. F. Heinz and D. Ricard and Y. R. Shen. *Phys. Rev. B*, 27(4):1965–1979.

Chen, S.-J., Chien, F. C., Lin, G. Y., and Lee, K. C. (2004). Enhancement of the resolution of surface plasmon resonance biosensors by control of the size and distribution of nanoparticles. *Opt. Lett.*, 29(12):1390–1392.

Citrin, D. S. (2004). Coherent excitation transport in metal-nanoparticle chains. *Nano Letters*, 4(9):1561–1565.

Citrin, D. S. (2005a). Plasmon-polariton transport in metal-nanoparticle chains embedded in a gain medium. *Opt. Lett.*, 31(1):98–100.

Citrin, D. S. (2005b). Plasmon polaritons in finite-length metal-nanoparticle chains: The role of chain length unravelled. *Nano Letters*, 5(5):985–989.

Cognet, L., Tardin, C., Boyer, D., Choquet, D., Tamarat, P., and Lounis, B. (2003). Single metallic nanoparticle imaging for protein detection in cells. *Proceedings of the National Academy of Sciences (USA)*, 100(20):11350–11355.

Coutaz, J. L., Neviere, M., Pic, E., and Reinisch, R. (1985). Experimental study of surface-enhanced second-harmonic generation on silver gratings. *Phys. Rev. B*, 32(4):2227–2232.

Craighead, H. G. and Niklasson, G. A. (1984). Characterization and optical properties of arrays of small gold particles. *Appl. Phys. Lett.*, 44(12):1134–1136.

Daniels, Jacquitta K. and Chumanov, George (2005). Nanoparticle-mirror sandwich substrates for surface-enhanced Raman scattering. *J. Phys. Chem. B*, 109:17936–17942.

Dawson, P., de Fornel, F., and Goudonnet, J-P. (1994). Imaging of surface plasmon propagation and edge interaction using a photon scanning tunneling microscope. *Phys. Rev. Lett.*, 72(18):2927–2930.

de Abajo, F. J. Garcia (2002). Light transmission through a single cylindrical hole in a metallic film. *Opt. Express*, 10(25):1475–1484.

de Abajo, F. J. Garcia and Sáenz, J. J. (2005). Electromagnetic surface modes in structured perfect-conductor surfaces. *Phys. Rev. Lett.*, 95:233901.

de Abajo, F. J. García, Sáenz, J. J., Campillo, I., and Dolado, J. S. (2006). Site and lattice resonances in metallic hole arrays. *Opt. Express*, 14(1):7–18.

Degiron, A. and Ebbesen, T. W. (2005). The role of localized surface plasmon modes in the enhanced transmission of periodic subwavelength apertures. *J. Opt. A:Pure Appl. Opt.*, 7:S90–S96.

Degiron, A., Lezec, H. J., Barnes, W. L., and Ebbesen, T. W. (2002). Effects of hole depth on enhanced light transmission through subwavelength hole arrays. *Appl. Phys. Lett.*, 81(23):4327–4329.

Degiron, A., Lezec, H. J., Yamamoto, N., and Ebbesen, T. W. (2004). Optical transmission properties of a single subwavelength aperture in a real metal. *Opt. Commun.*, 239:61–66.

Depine, Ricardo A. and Ledesma, Silvia (2004). Direct visualization of surface-plasmon bandgaps in the diffuse background of metallic gratings. *Opt. Lett.*, 29(19):2216–2218.

Dereux, A., Devaux, E., Weeber, J. C., Goudonnet, J. P., and Girard, C. (2001). Direct interpretation of near-field optical images. *J. Microscopy*, 202:320–331.

Devaux, Eloise, Ebbesen, Thomas W., Weeber, Jean-Claude, and Dereux, Alain (2003). Launching and decoupling surface plasmons via micro-gratings. *Appl. Phys. Lett.*, 83(24):4936–4938.

Dice, G. D., Mujumdar, S., and Elezzabi, A. Y. (2005). Plasmonically enhanced diffusive and subdiffusive metal nanoparticle-dye random laser. *Appl. Phys. Lett.*, 86:131105.

Dickson, Robert M. and Lyon, L. Andrew (2000). Unidirectional plasmon propagation in metallic nanowires. *J. Phys. Chem. B*, 104:6095–6098.

Diez, Antonio, Andrés, Miguel V., and Cruz, José L. (1999). Hybrid surface plasma modes in circular metal-coated tapered fibers. *J. Opt. Soc. Am. A*, 16(12):2978–2982.

Ditlbacher, H., Krenn, J. R., Félidj, N., Lamprecht, B., Schider, G., Salerno, M., Leitner, A., and Aussenegg, F. R. (2002a). Fluorescence imaging of surface plasmon fields. *Appl. Phys. Lett.*, 80(3):404–406.

Ditlbacher, H., Krenn, J. R., Hohenau, A., Leitner, A., and Aussenegg, F. R. (2003). Efficiency of local light-plasmon coupling. *Appl. Phys. Lett.*, 83(18):3665–3667.

Ditlbacher, H., Krenn, J. R., Schider, G., Leitner, A., and Aussenegg, F. R. (2002b). Two-dimensional optics with surface plasmon polaritons. *Appl. Phys. Lett.*, 81(10):1762–1764.

Ditlbacher, Harald, Hohenau, Andreas, Wagner, Dieter, Kreibig, Uwe, Rogers, Michael, Hofer, Ferdinand, Aussenegg, Franz R., and Krenn, Joachim R. (2005). Silver nanowires as surface plasmon resonators. *Phys. Rev. Lett.*, 95:257403.

Dragnea, Bogdan, Szarko, Jodi M., Kowarik, Stefan, Weimann, Thomas, Feldmann, Jochen, and Leone, Stephen R. (2003). Near-field surface plasmon excitation on structured gold films. *Nano Letters*, 3(1):3–7.

Drezet, A., Woehl, J. C., and Huant, S. (2001). Extension of Bethe's diffraction model to conical geometry: Application to near-field optics. *Europhysics Letters*, 54(6):736–740.

Drude, Paul (1900). Zur Elektronentheorie der Metalle. *Ann. Phys.*, 1:566–613.

Dulkeith, E., Morteani, A. C., Niedereichholz, T., Klar, T. A., Feldmann, J., , Levi, S. A., van Veggel, F. C. J. M., Neinhoudt, D. N., Moller, M., and Gittins, D. I. (2002). Fluorescence quenching of dye molecules near gold nanoparticles: Radiative and nonradiative effects. *Phys. Rev. Lett.*, 89(20):203002.

Dulkeith, E., Niedereichholz, T., Klar, T. A., Feldmann, J., von Plessen, G., Gittins, D. I., Mayya, K. S., and Caruso, F. (2004). Plasmon emission in photoexcited gold nanoparticles. *Phys. Rev. B*, 70:205424.

Eah, Sang-Kee, Jaeger, Heinrich M., Scherer, Norbert F., Wiederrecht, Gary P., and Lin, Xiao-Min (2005). Plasmon scattering from a single gold nanoparticle collected through an optical fiber. *Appl. Phys. Lett.*, 96:031902.

Ebbesen, T. W., Lezec, H. J, Ghaemi, H. F., Thio, T., and Wolff, P. A. (1998). Extraordinary optical transmission through sub-wavelength hole arrays. *Nature*, 931:667–669.

Economou, E. N. (1969). Surface plasmons in thin films. *Phys. Rev.*, 182(2):539–554.

El-Sayed, Ivan H., Huang, Xiaohua, and El-Sayed, Mostafa A. (2005). Surface plasmon resonance scattering and absorption of anti-EGFR antibody conjugated gold nanoparticles in cancer diagnostics: Applications in oral cancer. *Nano Letters*, 5(5):829–834.

Enoch, Stefan, Quidant, Romain, and Badenes, Goncal (2004). Optical sensing based on plasmon coupling in nanoparticle arrays. *Opt. Express*, 12(15):3422–3427.

Fang, Nicolas, Lee, Hyesog, Sun, Cheng, and Zhang, Xiang (2005). Sub-diffraction-limeted optical imaging with a silver superlens. *Science*, 308:534–537.

Fano, U. (1941). The theory of anomalous diffraction gratings and of quasi-stationary waves on metallic surfaces (Sommerfeld's waves). *J. Opt. Soc. Am.*, 31:213–222.

Farrer, Richard A., Butterfield, Francis L., Chen, Vincent W., and Fourkas, John T. (2005). Highly efficient multiphoton-absorption-induced luminescence from gold nanoparticles. *Nano Lett.*, 5(6):1139–1141.

Félidj, N., Aubard, J., Lévi, G., Krenn, J. R., Schider, G., Leitner, A., and Aussenegg, F. R. (2002). Enhanced substrate-induced coupling in two-dimensional gold nanoparticle arrays. *Phys. Rev. B*, 66:245407.

Félidj, N., Truong, S. Lau, Aubard, J., Lévi, G., Krenn, J. R., Hohenau, A., Leitner, A., and Aussenegg, F. R. (2004). Gold particle interaction in regular arrays probed by surface enhanced Raman scattering. *J. Chem. Phys.*, 120(15):7141–7146.

Ganeev, R. A., Baba, M., Ryasnyansky, A. I., Suzuki, M., and Kuroda, H. (2004). Characterization of optical and nonlinear optical properties of silver nanoparticles prepared by laser ablation in various liquids. *Opt. Commun.*, 240:437–448.

García-Vidal, F. J., Lezec, H. J., Ebbesen, T. W., and Martín-Moreno, L. (2003a). Multiple paths to enhance optical transmission through a single subwavelength slit. *Phys. Rev. Lett.*, 90(21):213901.

García-Vidal, F. J., Martín-Moreno, L., Lezec, H. J., and Ebbesen, T. W. (2003b). Focusing light with a single subwavelength aperture flanked by surface corrugations. *Appl. Phys. Lett.*, 83(22):4500–4502.

García-Vidal, F. J., Martín-Moreno, L., and Pendry, J. B. (2005a). Surfaces with holes in them: new plasmonic metamaterials. *J. Opt. A: Pure Appl. Opt.*, 7:S97–S101.

García-Vidal, F. J., Moreno, Esteban, Porto, J. A., and Martín-Moreno, L. (2005b). Transmission of light through a single rectangular hole. *Phys. Rev. Lett.*, 95:103901.

García-Vidal, F. J. and Pendry, J. B. (1996). Collective theory for surface enhanced Raman scattering. *Phys. Rev. Lett.*, 77(6):1163–1166.

Gersten, Joel and Nitzan, Abraham (1980). Electromagnetic theory of enhanced Raman scattering by molecules adsorbed on rough surfaces. *J. Chem. Phys.*, 73(7):3023–3037.

Ghaemi, H. F., Thio, Tineke, Grupp, D. E., Ebbesen, T. W., and Lezec, H. J. (1998). Surface plasmon enhance optical transmission through subwavelength holes. *Phys. Rev. B*, 58(11):6779–6782.

Giannattasio, Armando and Barnes, William L. (2005). Direct observation of surface plasmon-polariton dispersion. *Opt. Express*, 13(2):428–434.

Giannattasio, Armando, Hooper, Ian R., and Barnes, William L. (2004). Transmission of light through thin silver films via surface plasmon-polaritons. *Opt. Express*, 12(24):5881–5886.

Girard, Christian and Quidant, Romain (2004). Near-field optical transmittance of metal particle chain waveguides. *Opt. Express*, 12(25):6141–6146.

Gómez-Rivas, J., Kuttge, M., Bolivar, P. Haring, Kurz, H., and Sánchez-Gill, J. A. (2004). Propagation of surface plasmon polaritons on semiconductor gratings. *Phys. Rev. Lett.*, 93:256804.

Gómez-Rivas, J., Kuttge, M., Kurz, H., Bolivar, P. Haring, and Sánchez-Gill, J. A. (2006). Low-frequency active surface plasmon optics on semiconductors. *Appl. Phys. Lett.*, 88:082106.

Gordon, Reuven and Brolo, Alexandre G. (2005). Increased cut-off wavelength for a subwavelength hole in a real metal. *Opt. Express*, 13(6):1933–1938.

Goubau, Georg (1950). Surface waves and their application to transmission lines. *J. Appl. Phys.*, 21:1119–1128.

Grand, J., de la Chapelle, M. Lamy, Bijeon, J.-L., Adam, P.-M., and Royer, P. (2005). Role of localized surface plasmons in surface-enhanced Raman scattering of shape-controlled metallic particles in regular arrays. *Phys. Rev. B*, 72:033407.

Grigorenko, A. N., Geim, A. K., Gleeson, H. F., Zhang, Y., Firsov, A. A., Khrushchev, I. Y., and Petrovic, J. (2005). Nanofabricated media with negative permeability at visible frequencies. *Nature*, 438:335–338.

Grupp, Daniel E., Lezec, Henri J., Thio, Tineke, and Ebbesen, Thomas W. (1999). Beyond the Bethe limit: Tunable enhanced light transmission through a single sub-wavelength aperture. *Advanced Materials*, 11(10):860–862.

Haes, Amanda J., Hall, W. Paige, Chang, Lei, Klein, William L., and van Dyne, Richard P. (2004). A localized surface plasmon resonance biosensor: First steps toward and assay for Alzheimer's disease. *Nano Letters*, 4(6):1029–1034.

Hartschuh, Achim, Sánchez, Erik J., Xie, X. Sunney, and Novotny, Lukas (2003). High-resolution near-field Raman microscopy of single-walled carbon nanotubes. *Phys. Rev. Lett.*, 90(9):095503.

Haus, Hermann A. (1984). *Waves and Fields in Optoelectronics*. Prentice-Hall, Englewood Cliffs, New Jersey 07632, first edition.

Hayazawa, Norihiko, Saito, Yuika, and Kawata, Satoshi (2004). Detection and characterization of longitudinal field for tip-enhanced Raman spectroscopy. *Appl. Phys. Lett.*, 85(25):6239–6241.

Haynes, Christy L., McFarland, Adam D., Zhao, LinLin, Duyne, Richard P. Van, Schatz, George C., Gunnarsson, Linda, Prikulis, Juris, Kasemo, Bengt, and Käll, Mikael (2003). Nanoparticle optics: The importance of radiative dipole coupling in two-dimensional nanoparticle arrays. *J. Phys. Chem. B*, 107:7337–7342.

Hecht, B., Bielefeld, H., Novotny, L., Inouye, Y., and Pohl, D. W. (1996). Local excitation, scattering, and interference of surface plasmons. *Phys. Rev. Lett.*, 77(9):1889–1892.

Heilweil, E. J. and Hochstrasser, R. M. (1985). Nonlinear spectroscopy and picosecond transient grating study of colloidal gold. *J. Chem. Phys.*, 82(9):4762–4770.

Hibbins, Alastair P., Evans, Benjamin R., and Sambles, J. Roy (2005). Experimental verification of designer surface plasmons. *Science*, 308:670–672.

Hibbins, Alastair P., Lockyear, Matthew J., Hooper, Ian R., and Sambles, J. Roy (2006). Waveguide arrays as plasmonic metamaterials: transmission below cutoff. *Phys. Rev. Lett.*, 96:073904.

Hicks, Erin M., Zou, Shengli, Schatz, George C., Spears, Kenneth G., Duyne, Richard P. Van, Gunnarsson, Linda, Rindzevicius, Tomas, Kasemo, Bengt, and Käll, Mikael (2005). Controlling plasmon line shapes through diffractive coupling in linear arrays of cylindrical nanoparticles fabricated by electron beam lithography. *Nano Lett.*, 5(6):1065–1070.

Hillenbrand, R., Taubner, T., and Keilmann, F. (2002). Phonon-enhanced light-matter interaction at the nanometre scale. *Nature*, 418:159–162.

Hinds, E.A. (1994). *Pertubative cavity quantum electrodynamics*, pages 1–56. Academic Press, Boston.

Hirsch, L. R., Stafford, R. J., Bankson, J. A., Sershen, S. R., Rivera, B., Price, R. E., Hazle, J. D., Halas, N. J., and West, J. L. (2003). Nanoshell-mediated near-infrared thermal therapy of tumors under magnetic resonance guiding. *Proc. Nat. Acad. Sci.*, 100(23):13549–13554.

Hochberg, Michael, Baehr-Jones, Tom, Walker, Chris, and Scherer, Axel (1985). Integrated plasmon and dielectric waveguides. *Opt. Express*, 12(22):5481–5486.

Hohenau, A., Krenn, J. R., Schider, G., Ditlbacher, H., Leitner, A., Aussenegg, F. R., and Schaich, W. L. (2005a). Optical near-field of multipolar plasmons of rod-shaped gold nanoparticles. *Europhys. Lett.*, 69(4):538–543.

Hohenau, Andreas, Krenn, Joachim R., Stepanov, Andrey L., Drezet, Aurelien, Ditlbacher, Harald, Steinberger, Bernhard, Leitner, Alfred, and Aussenegg, Franz R. (2005b). Dielectric optical elements for surface plasmons. *Opt. Lett.*, 30(8):893–895.

Homola, Jirí, Slavik, Radan, and Ctyroky, Jiri (1997). Interaction between fiber modes and surface plasmon waves: spectral properties. *Opt. Lett.*, 22(18):1403–1405.

Homola, Jirí, Yee, Sinclair S., and Gauglitz, Gunter (1999). Surface plasmon reonance sensors: review. *Sensors and Actuators B*, 54:3–15.

Hooper, I. R. and Sambles, J. R. (2002). Dispersion of surface plasmon polaritons on short-pitch metal gratings. *Phys. Rev. B*, 65:165432.

Hooper, I. R. and Sambles, J. R. (2004a). Coupled surface plasmon polaritons on thin metal slabs corrugated on both surfaces. *Phys. Rev. B*, 70:045421.

Hooper, I. R. and Sambles, J. R. (2004b). Differential ellipsometric surface plasmon resonance sensors with liquid crystal polarization modulators. *Appl. Phys. Lett.*, 85(15):3017–1019.

Hövel, H., Fritz, S., Hilger, A., Kreibig, U., and Vollmer, M. (1993). Width of cluster plasmon resonances: Bulk dielectric functions and chemical interface damping. *Phys. Rev. B*, 48(24):18178–18188.

Huber, A., Ocelic, N., Kazantsev, D., and Hillenbrand, R. (2005). Near-field imaging of mid-infrared surface phonon polariton propagation. *Appl. Phys. Lett.*, 87:081103.

Illinskii, Yu. A. and Keldysh, L. V. (1994). *Electromagnetic response of material media*. Plenum Press, New York, NY, first edition.

Imura, Kohei, Nagahara, Tetsuhiko, and Okamoto, Hiromi (2005). Near-field optical imaging of plasmon modes in gold nanorods. *J. Chem. Phys.*, 122:154701.

Jackson, John D. (1999). *Classical Electrodynamics*. John Wiley & Sons, Inc., New York, NY, 3rd edition.

Jeon, Tae-In and Grischkowsky, D. (2006). THz Zenneck surface wave (THz surface plasmon) propagation on a metal sheet. *Appl. Phys. Lett.*, 88:061113.

Jeon, Tae-In, Zhang, Jiangquan, and Grischkowsky, D. (2005). THz Sommerfeld wave propagation on a single metal wire. *Appl. Phys. Lett.*, 86:161904.

Jette-Charbonneau, Stephanie, Charbonneau, Robert, Lahoud, Nancy, Mattiussi, Greg, and Berrini, Pierre (2005). Demonstration of Bragg gratings based on long-ranging surface plasmon polariton waveguides. *Opt. Express*, 13(12):4674–4682.

Johannsson, Peter, Xu, Hongxing, and Käll, Mikael (2005). Surface-enhanced Raman scattering and fluorescence near metal nanoparticles. *Phys. Rev. B*, 72:035427.

Johnson, P. B. and Christy, R. W. (1972). Optical constants of the noble metals. *Phys. Rev. B*, 6(12):4370–4379.

Kashiwa, Tatsuya and Fukai, Ichiro (1990). A treatment by the FD-TD method of the dispersive characteristics associated with electronic polarization. *Microwave and Optical Technology Letters*, 3(6):203–205.

Kerker, Milton, Wang, Dau-Sing, and Chew, H. (1980). Surface enhanced Raman scattering (SERS) by molecules adsorbed at spherical particles: errata. *Appl. Opt.*, 19(24):4159–4147.

Kim, Yoon-Chang, Peng, Wei, Banerji, Soame, and Booksh, Karl S. (2005). Tapered fiber optic surface plasmon resonance sensor for analyses of vapor and liquid phases. *Opt. Lett.*, 30(17):2218–2220.

Kittel, Charles (1996). *Introduction to Solid State Physics*. John Wiley & Sons, Inc., New York, NY, seventh edition edition.

Klar, T., Perner, M., Grosse, S., von Plessen, G., Spirkl, W., and Feldmann, J. (1998). Surface-plasmon resonances in single metallic nanoparticles. *Phys. Rev. Lett.*, 80(19):4249–4252.

Kneipp, K., Kneipp, H., Itzkan, I., Dasari, R. R., and Feld, M. S. (2002). Surface enhanced Raman scattering and biophysics. *J. Phys. Cond. Mat.*, 14:R597–R624.

Kneipp, K., Wang, Y., Kneipp, H., Perelman, L. T., Itzkan, I., Dasari, R. R., and Feld, M. S. (1997). Single molecule detection using surface-enhanced Raman scattering (SERS). *Phys. Rev. Lett.*, 78(9):1667.

Kokkinakis, Th. and Alexopoulos, K. (1972). Observation of radiative decay of surface plasmons in small silver particles. *Phys. Rev. Lett.*, 28(25):1632–1634.

Kreibig, U. and Vollmer, M. (1995). *Optical properties of metal clusters*. Springer, Berlin.

Krenn, J. R., Dereux, A., Weeber, J. C., Bourillot, E., Lacroute, Y., Goudonnet, J. P., Schider, G., Gotschy, W., Leitner, A., Aussenegg, F. R., and Girard, C. (1999). Squeezing the optical near-field zone by plasmon coupling of metallic nanoparticles. *Phys. Rev. Lett.*, 82(12):2590–2593.

Krenn, J. R., Lamprecht, B., Ditlbacher, H., Schider, G., Salerno, M., Leitner, A., and Aussenegg, F. R. (2002). Non-diffraction-limited light transport by gold nanowires. *Europhys. Lett.*, 60(5):663–669.

Krenn, J. R., Salerno, M., Félidj, N., Lamprecht, B., Schider, G., Leitner, A., Aussenegg, F. R., Weeber, J. C., Dereux, A., and Goudonnet, J. P. (2001). Light field propagation by metal micro- and nanostructures. *J. Microscopy*, 202:122–128.

Krenn, J. R., Schider, G., Rechenberger, W., Lamprecht, B., Leitner, A., Aussenegg, F. R., and Weeber, J. C. (2000). Design of multipolar plasmon excitations in silver nanoparticles. *Appl. Phys. Lett.*, 77(21):3379–3381.

Kretschmann, E. (1971). Die Bestimmung optischer Konstanten von Metallen durch Anregung von Oberflächenplasmaschwingungen. *Z. Physik*, 241:313–324.

Kretschmann, E. and Raether, H. (1968). Radiative decay of non-radiative surface plasmons excited by light. *Z. Naturforschung*, 23A:2135–2136.

Kuwata, Hitoshi, Tamaru, Hiroharu, Esumi, Kunio, and Miyano, Kenjiro (2003). Resonant light scattering from metal nanoparticles: Practical analysis beyond Rayleigh approximation. *Appl. Phys. Lett.*, 83(22):4625–2627.

Lamprecht, B., Krenn, J. R., Leitner, A., and Aussenegg, F. R. (1999). Resonant and off-resonant light-driven plasmons in metal nanoparticles studied by femtosecond-resolution third-harmonic generation. *Phys. Rev. Lett.*, 83(21):4421–4424.

Lamprecht, B., Krenn, J. R., Schider, G., Ditlbacher, H., Salerno, M., Félidj, N., Leitner, A., Aussenegg, F. R., and Weeber, J. C. (2001). Surface plasmon propagation in microscale metal stripes. *Appl. Phys. Lett.*, 79(1):51–53.

Lamprecht, B., Schider, G., Ditlbacher, R. T. Lechner H., Krenn, J. R., Leitner, A., Aussenegg, F. R., and Weeber, J. C. (2000). Metal nanoparticle gratings: Influence of dipolar particle interaction on the plasmon resonance. *Phys. Rev. Lett.*, 84(20):4721–4724.

Larkin, Ivan A., Stockman, Mark I., Achermann, Marc, and Klimov, Victor I. (2004). Dipolar emitters at nanoscale proximity of metal surfaces: Giant enhancement of relaxation in microscopic theory. *Phys. Rev. B*, 69:121403(R).

Laurent, G., Félidj, N., Truong, S. Lau, Aubard, J., Levi, G., Krenn, J. R., Hohenau, A., Leitner, A., and Aussenegg, F. R. (2005a). Imaging surface plasmon of gold nanoparticle arrays by far-field Raman scattering. *Nano Letters*, 5(2):253–258.

Laurent, G., Félidj, N., Truong, S. Lau, Aubard, J., Levi, G., Krenn, J. R., Hohenau, A., Leitner, G. Schider A., and Aussenegg, F. R. (2005b). Evidence of multipolar excitations in surface enhanced Raman scattering. *Phys. Rev. B*, 71:045430.

Lawandy, L. M. (2004). Localized surface plasmon singularities in amplifying media. *Appl. Phys. Lett.*, 85(21):5040–5042.

Leosson, K., Nikolajsen, T., Boltasseva, A., and Bozhevolnyi, S. I. (2006). Long-range surface plasmon polariton nanowire waveguides for device applications. *Opt. Express*, 14(1):314–319.

Lezec, H. J., Degiron, A., Devaux, E., Linke, R. A., Martin-Moreno, L., Garcia-Vidal, F. J, and Ebbesen, T. W. (2002). Beaming light from a subwavelength aperture. *Science*, 297:820–822.

Li, Kuiru, Stockman, Martin I., and Bergman, David J. (2003). Self-similar chain of metal nanospheres as an efficient nanolens. *Phys. Rev. Lett.*, 91(22):227402.

Liao, P. F. and Wokaun, A. (1982). Lightning rod effect in surface enhanced Raman scattering. *J. Chem. Phys.*, 76(1):751–752.

Lin, Haohao, Mock, Jack, Smith, David, Gao, Ting, and Sailor, Michael J. (2004). Surface-enhanced Raman scattering from silver-plated porous silicon. *J. Phys. Chem. B*, 108:11654–11659.

Link, Stephan and El-Sayed, Mostafa A. (2000). Shape and size dependence of radiative, non-radiative and photothermal properties of gold nanocrystals. *Int. Reviews in Physical Chemistry*, 19(3):409–453.

Lippitz, Markus, van Dijk, Meindert A., and Orrit, Michel (2005). Third-harmonic generation from single gold nanoparticles. *Nano Lett.*, 5(4):799–802.

Liu, Hongwen, Ie, Yatuka, Yoshinobu, Tasuo, Aso, Yoshio, Iwasaki, Hiroshi, and Nishitani, Ryusuke (2006). Plasmon-enhanced molecular fluorescence from an organic film in a tunnel junction. *Appl. Phys. Lett.*, 88:061901.

Liu, Zahowei, Steele, Jennifer M., Srituravanich, Werayut, Pikus, Yuri, Sun, Cheng, and Zhang, Xiang (2005). Focusing surface plasmons with a plasmonic lens. *Nano Letters*, 5(9):1726–1729.

Loudon, R. (1970). The propagation of electromagnetic energy through an absorbing dielectric. *J. Phys. A*, 3:233–245.

Lu, H. Peter (2005). Site-specific Raman spectroscopy and chemical dynamics of nanoscale interstitial systems. *J. Phys.: Condens. Matter*, 17:R333–R355.

Lu, Yu, Liu, Gang L., Kim, Jaeyoun, Mejia, Yara X., and Lee, Luke P. (2005). Nanophotonic crescent moon structures with sharp edge for ultrasensitive biomolecular detection by local electromagnetic field enhancement effect. *Nano Letters*, 5(1):119–124.

Luo, Xiangang and Ishihara, Teruya (2004). Surface plasmon resonant interference nanolithography technique. *Appl. Phys. Lett.*, 84(23):4780–4782.

Maier, Stefan A. (2006a). Gain-assisted propagation of electromagnetic energy in subwavelength surface plasmon polariton gap waveguides. *Opt. Commun.*, 258:295–299.

Maier, Stefan A. (2006b). Plasmonic field enhancement and SERS in the effective mode volume picture. *Opt. Express*, 14(5):1957–1964.

Maier, Stefan A., Barclay, Paul E., Johnson, Thomas J., Friedman, Michelle D., and Painter, Oskar (2004). Low-loss fiber accessible plasmon waveguides for planar energy guiding and sensing. *Appl. Phys. Lett.*, 84(20):3990–3992.

Maier, Stefan A., Brongersma, Mark L., Kik, Pieter G., and Atwater, Harry A. (2002a). Observation of near-field coupling in metal nanoparticle chains using far-field polarization spectroscopy. *Phys. Rev. B*, 65:193408.

Maier, Stefan A., Brongersma, Mark L., Kik, Pieter G., Meltzer, Sheffer, Requicha, Ari A. G., and Atwater, Harry A. (2001). Plasmonics - a route to nanoscale optical devices. *Adv. Mat.*, 13(19):1501.

Maier, Stefan A., Friedman, Michelle D., Barclay, Paul E., and Painter, Oskar (2005). Experimental demonstration of fiber-accessible metal nanoparticle plasmon waveguides for planar energy guiding and sensing. *Appl. Phys. Lett.*, 86:071103.

Maier, Stefan A., Kik, Pieter G., and Atwater, Harry A. (2002b). Observation of coupled plasmon-polariton modes in Au nanoparticle chain waveguides of different lengths: Estimation of waveguide loss. *Appl. Phys. Lett.*, 81:1714–1716.

Maier, Stefan A., Kik, Pieter G., and Atwater, Harry A. (2003a). Optical pulse propagation in metal nanoparticle chain waveguides. *Phys. Rev. B*, 67:205402.

Maier, Stefan A., Kik, Pieter G., Atwater, Harry A., Meltzer, Sheffer, Harel, Elad, Koel, Bruce E., and Requicha, Ari A. G. (2003b). Local detection of electromagnetic energy transport below the diffraction limit in metal nanoparticle plasmon waveguides. *Nat. Mat.*, 2(4):229–232.

Marder, Michael P. (2000). *Condensed Matter Physics*. John Wiley & Sons, Inc., New York, NY.

Markel, V. A., Shalaev, V. M., Zhang, P., Huynh, W., Tay, L., Haslett, T. L., and Moskovits, M. (1999). Near-field optical spectroscopy of individual surface-plasmon modes in colloid clusters. *Phys. Rev. B*, 59(16):10903–10909.

Marquart, Carsten, Bozhevolnyi, Sergey I., and Leosson, Kristjan (2005). Near-field imaging of surface plasmon-polariton guiding in band gap structures at telecom wavelengths. *Opt. Express*, 13(9):3303–3309.

Marquier, F., Greffet, J.-J., Collin, S., Pardo, F., and Pelouard, J. L. (2005). Resonant transmission through a metallic films due to coupled modes. *Opt. Express*, 13(1):70–76.

Marti, O., Bielefeld, H., Hecht, B., Herminghaus, S., Leiderer, P., and Mlynek, J. (1993). Mear-field optical measurement of the surface plasmon field. *Opt. Commun.*, 96(4–6):225–228.

Martín-Moreno, L., Garcia-Vidal, F. J., Lezec, H. J., Degiron, A., and Ebbesen, T. W. (2003). Theory of highly directional emission from a single subwavelength aperture surrounded by surface corrugations. *Phys. Rev. Lett.*, 90(16):167401.

Matsui, Tatsunosuke, Vardeny, Z. Valy, Agrawal, Amit, Nahata, Ajay, and Menon, Reghu (2006). Resonantly-enhanced transmission through a periodic array of subwavelength apertures in heavily-doped conducting polymer films. *Appl. Phys. Lett.*, 88:071101.

Meier, M. and Wokaun, A. (1983). Enhanced fields on large metal particles: dynamic depolarization. *Opt. Lett.*, 8(11):581–583.

Melville, David O. S. and Blaikie, Richard J. (2005). Super-resolution imaging through a planar silver layer. *Opt. Express*, 13(6):2127–2134.

Melville, David O. S. and Blaikie, Richard J. (2006). Super-resolution imaging through a planar silver layer. *J. Opt. Soc. Am. B*, 23(3):461–467.

Mie, Gustav (1908). Beiträge zur Optik trüber Medien, speaiell kolloidaler Metallösungen" . *Ann. Phys.*, 25:377.

Mikhailovsky, A. A., Petruska, M. A., Li, Kuiru, Stockman, M. I., and Klimov, V. I. (2004). Phase-sensitive spectroscopy of surface plasmons in individual metal nanostructures" . *Phys. Rev. B*, 69:085401.

Mikhailovsky, A. A., Petruska, M. A., Stockman, M. I., and Klimov, V. I. (2003). Broadband near-field interference spectroscopy of metal nanoparticles using a femtosecond white-light continuum. *Opt. Lett.*, 28(18):1686–1688.

Milner, R. G. and Richards, D. (2001). The role of tip plasmons in near-field Raman microscopy. *J. Microscopy*, 202:66–71.

Mitsui, Keita, Handa, Yoichiro, and Kajikawa, Kotaro (2004). Optical fiber affinity biosensor based on localized surface plasmon resonance" . *Appl. Phys. Lett.*, 85(18):4231–4233.

Mock, J. J., Barbic, M., Smith, D. R., Schultz, D. A., and Schultz, S. (2002a). Shape effects in plasmon resonance of individual colloidal silver nanoparticles". *J. Chem. Phys.*, 116(15):6755–6759.

Mock, J. J., Oldenburg, S. J., Smith, D. R., Schultz, D. A., and Schultz, S. (2002b). Composite plasmon resonant nanowires". *Nano Letters*, 2(5):465–469.

Mock, Jack J., Smith, David R., and Schultz, Sheldon (2003). Local refractive index dependence of plasmon resonance spectra from individual nanoparticles". *Nano Letters*, 3(4):485–491.

Monzon-Hernandez, David, Villatoro, Joel, Talavera, Dimas, and Luna-Moreno, Donato (2004). Optical-fiber surface-plasmon resonance sensor with multiple resonance peaks. *Appl. Opt.*, 43(6):1216–1220.

Mooradian, A. (1969). Photoluminescence of metals. *Phys. Rev. Lett.*, 22(5):185–187.

Moreno, Estaban, Fernández-Domínguez, A. I., Cirac, J. Ignacia, Garía-Vidal, F. J., and Martín-Moreno, L. (2005). Resonant transmission of cold atoms through subwavelength apertures. *Phys. Rev. Lett.*, 95:170406.

Moser, H. O., Casse, B. D. F., Wilhelmi, O., and Shaw, B. T. (2005). Terahertz response of a microfabricated rod-split-ring-resonator electromagnetic metamaterial. *Phys. Rev. Lett.*, 94:063901.

Moskovits, Martin (1985). Surface-enhanced spectroscopy. *Reviews of Modern Physics*, 57(3): 783–826.

Nenninger, G. G., Tobiska, P., Homola, J., and Yee, S. S. (2001). Long-range surface plasmons for high-resolution surface plasmon resonance sensors. *Sensors and Actuators B*, 74:145–151.

Nezhad, Maziar P., Tetz, Kevin, and Fainman, Yeshaiahu (2004). Gain assisted propagation of surface plasmon polaritons on planar metallic waveguides. *Opt. Express*, 12(17):4072.

Nie, S. M. and Emery, S. R. (1997). Probing single molecules and single nanoparticles by surface-enhanced Raman scattering. *Science*, 275(5303):1102.

Nienhuys, Han-Kwang and Sundström, Villy (2005). Influence of plasmons on terahertz conductivity measurements. *Appl. Phys. Lett.*, 87:012101.

Nikolajsen, Thomas, Leosson, Kristjan, and Bozhevolnyi, Sergey I. (2004a). Surface plasmon polariton based modulators and switches operating at telecom wavelengths. *Appl. Phys. Lett.*, 85(24):5833–5835.

Nikolajsen, Thomas, Leosson, Kristjan, and Bozhevolnyi, Sergey I. (2004b). Surface plasmon polariton based modulators and switches operating at telecom wavelengths. *Appl. Phys. Lett.*, 85(24):5833–5835.

Nomura, Wataru, Ohtsu, Motoichi, and Yatsui, Takashi (2005). Nanodot coupler with a surface plasmon polariton condensor for optical far/near-field conversion. *Appl. Phys. Lett.*, 86:181108.

Nordlander, P., Oubre, C., Prodan, E., Li, K., and Stockman, M. I. (2004). Plasmon hybridization in nanoparticle dimers. *Nano Lett.*, 4(5):899–903.

Novikov, I. V. and Maradudin, A. A. (2002). Channel polaritons. *Phys. Rev. B*, 66:035403.

Ocelic, N. and Hillenbrand, R. (2004). Subwavelength-scale tailoring of surface phonon polaritons by focused ion-beam implantation. *Nat. Mat.*, 3:606–609.

Offerhaus, H. L., van den Bergen, B., Escalante, M., Segerink, F. B., Korterik, J. P., and van Hulst, N. F. (2005). Creating focues plasmons by noncollinear phasematching on functional gratings. *Nano Lett.*, 5(11):2144–2148.

Olkkonen, Juuso, Kataja, Kari, and Howe, Dennis G. (2005). Light transmission through a high index dielectric-filled sub-wavelength hole in a metal film. *Nature*, 432:376–379.

Ordal, M. A., Long, L. L., Bell, R. J., Bell, R. R., Alexander, R. W., and Ward, C. A. (1983). Optical properties of the metals Al, Co, Cu, Au, Fe, Pb, Ni, Pd, Pt, Ag, Ti, and W in the infrared and far infrared. *Appl. Opt.*, 22(7):1099–1119.

Otto, A. (1968). Excitation of nonradiative surface plasma waves in silver by the method of frustrated total reflection. *Z. Physik*, 216:398–410.

Oubre, Chris and Nordlander, Peter (2004). Optical properties of metallodielectric nanostructures calculated using the finite difference time domain method. *J. Phys. Chem. B*, 108:17740–17747.

Park, Sung Yong and Stroud, David (2004). Surface-plasmon dispersion relations in chains of metallic nanoparticles: An exact quasistatic solution. *Phys. Rev. B*, 69:125418.

Park, Suntak, Lee, Gwansu, Song, Seok Ho, Oh, Cha Hwan, and Kim, Phill Soo (2003). Resonant coupling of surface plasmons to radiation modes by use of dielectric gratings. *Opt. Lett.*, 28(20):1870–1872.

Passian, A., Lereu, A. L., Wig, A., Meriaudeau, F., Thundat, T., and Ferrell, T. L. (2005). Imaging standing surface plasmons by photon tunneling. *Phys. Rev. B*, 71:165418.

Passian, A., Wig, A., Lereu, A. L., Meriaudeau, F., Thundat, T., and Ferrell, T. L. (2004). Photon tunneling via surface plasmon coupling. *Appl. Phys. Lett.*, 85(16):3420–3422.

Pendry, J. B. (2000). Negative refraction makes a perfect lens. *Phys. Rev. Lett.*, 85(18):3966–3969.

Pendry, J. B., Holden, A. J., Robbins, D. J., and Stewart, W. J. (1999). Magnetism from conductors and enhanced nonlinear phenomena. *IEEE Trans. Microwave Theory Tech.*, 47(11):2075–2084.

Pendry, J. B., Holden, A. J., Stewart, W. J., and Youngs, I. (1996). Extremely low frequency plasmons in metallic mesostructures. *Phys. Rev. Lett.*, 76(25):4773–4776.

Pendry, J. B., Martin-Moreno, L., and Garcia-Vidal, F. J. (2004). Mimicking surface plasmons with structured surfaces. *Science*, 305:847–848.

Pettinger, Bruno, Ren, Bin, Picardi, Gennaro, Schuster, Rolf, and Ertl, Gerhard (2004). Nanoscale probing of adsorbed species by tip-enhanced Raman spectroscopy. *Phys. Rev. Lett.*, 92(9): 096101.

Pettit, R. B., Silcox, J., and Vincent, R. (1975). Measurement of surface-plasmon dispersion in oxidized aluminum films. *Phys. Rev. B*, 11(8):3116–3123.

Pile, D. F. P. and Gramotnev, D. K. (2004). Channel plasmon-polariton in a triangular groove on a metal surface. *Opt. Lett.*, 29(10):1069.

Pile, D. F. P., Ogawa, T., Gramotnev, D. K., Matsuzaki, Y., Vernon, K. C., Yamaguchi, K., Okarnoto, T., Haraguchi, M., and Fukui, M. (2005). Two-dimensionally localized modes of a nanoscale gap plasmon waveguide. *Appl. Phys. Lett.*, 87:261114.

Porto, J. A., Garcia-Vidal, F. J., and Pendry, J. B. (1999). Transmission resonances on metallic gratings with very narrow slits. *Phys. Rev. Lett.*, 83(14):2845–2848.

Porto, J. A., Martín-Moreno, L., and García-Vidal, F. J. (2004). Optical bistability in subwavelength slit apertures containing nonlinear media. *Phys. Rev. B*, 70:081402(R).

Powell, C. J. and Swan, J. B. (1960). Effect of oxidation on the characteristic loss spectra of aluminum and magnesium. *Phys. Rev.*, 118(3):640–643.

Prade, B. and Vinet, J. Y. (1994). Guided optical waves in fibers with negative dielectric constant. *J. Lightwave Tech.*, 12(1):6–18.

Prade, B., Vinet, J. Y., and Mysyrowicz, A. (1991). Guided optical waves in planar heterostructures with negative dielectric constant. *Phys. Rev. B*, 44(24):13556–13572.

Prikulis, Juris, Hanarp, Per, Olofsson, Linda, Sutherland, Duncan, and Käll, Mikael (2004). Optical spectroscopy of nanometric holes in thin gold films. *Nano Lett.*, 4(6):1003–1007.

Prodan, E. and Nordlander, P. (2003). Structural tunability of the plasmon resonances in metallic nanoshells. *Nano Lett.*, 3(4):543–547.

Prodan, E., Nordlander, P., and Halas, N. J. (2003a). Electronic structure and optical properties of gold nanoshells. *Nano Lett.*, 3(10):1411–1415.

Prodan, E., Radloff, C., Halas, N. J., and Nordlander, P. (2003b). A hybridization model for the plasmon response of complex nanostructures. *Science*, 203:419–422.

Qiu, Min (2005). Photonic band structure for surface waves on structured metal surfaces. *Opt. Express*, 13(19):7583–7588.

Quail, J. C., Rako, J. G., and Simon, H. J. (1983). Long-range surface-plasmon modes in silver and aluminum films. *Opt. Lett.*, 8(7):377–379.

Quinten, M., Leitner, A., Krenn, J. R., and Aussenegg, F. R. (1998). Electromagnetic energy transport via linear chains of silver nanoparticles. *Opt. Lett.*, 23(17):1331–1333.

Quinten, Michael and Kreibig, Uwe (1993). Absorption and elastic scattering of light by particle aggregates. *Appl. Opt.*, 32(30):6173–6182.

Raether, Heinz (1988). *Surface Plasmons*, volume 111 of *Springer-Verlag Tracts in Modern Physics*. Springer-Verlag, New York.

Raschke, G., Brogl, S., Susha, A. S., Rogach, A. L., Klar, T. A., Feldmann, J., Fieres, B., Petkov, N., Bein, T., Nichtl, A., and Kurzinger, K. (2004). Gold nanoshells improve single nanoparticle molecular sensors. *Nano Letters*, 4(10):1853–1857.

Raschke, G., Kowarik, S., Franzl, T., Soennichsen, C., Klar, T. A., and Feldmann, J. (2003). Biomolecular recognition based on single gold nanoparticle light scattering. *Nano Letters*, 3(7):935–938.

Ren, Bin, Lin, Xu-Feng, Yang, Zhi-Lin, Liu, Guo-Kun, Aroca, Rocardo F., Mao, Bing-Wei, and Tian, Zhong-Qun (2003). Surface-enhanced Raman scattering in the ultraviolet spectral region: UV-SERS on Rhodium and Ruthenium electrodes. *J. Am. Chem. Soc.*, 125:9598–9599.

Renger, Jan, Grafström, Stefan, Eng, Lukas M., and Hillenbrand, Rainer (2005). Resonant light scattering by near-field-induced phonon polaritons. *Phys. Rev. B*, 71:075410.

Rigneault, Hervé, Capoulade, Jérémie, Dintinger, José, Wenger, Jérôme, Bonod, Nicolas, Popov, Evgeni, Ebbesen, Thomas W., and Lenne, Pierre-François (2005). Enhancement of single-molecule fluorescence detection in subwavelength apertures. *Phys. Rev. Lett.*, 95:117401.

Rindzevicius, Tomas, Alaverdyan, Yury, Dahlin, Andreas, Hook, Fredrik, Sutherland, Duncan S., and Käll, Mikael (2005). Plasmonic sensing characteristics of single nanometric holes. *Nano Lett.*, 5(11):2335–2339.

Ritchie, R. H. (1957). Plasma losses by fast electrons in thin films. *Phys. Rev.*, 106(5):874–881.

Ritchie, R. H., Arakawa, E. T., Cowan, J. J., and Hamm, R. N. (1968). Surface-plasmon resonance effect in grating diffraction. *Phys. Rev. Lett.*, 21(22):1530–1533.

Roberts, A. (1987). Electromagnetic theory of diffraction by a circular aperture in a thick, perfectly conducting screen. *J. Opt. Soc. Am. A*, 4(10):1970–1983.

Rudnick, Joseph and Stern, E. A. (1971). Second-harmonic generation from metal surfaces. *Phys. Rev. B*, 4(12):4274–4290.

Ruppin, R. (2002). Electromagnetic energy density in a dispersive and absorptive material. *Phys. Lett. A*, 299:309–312.

Ruppin, R. (2005). Effect of non-locality on nanofocusing of surface plasmon field intensity in a conical tip. *Physics Letters A*, 340:299–302.

Saleh, Bahaa E. A. and Teich, Malvin Carl (1991). *Fundamentals of Photonics*. John Wiley & Sons, Inc., New York, NY.

Sarid, Dror (1981). Long-range surface-plasma waves on very thin metal films. *Phys. Rev. Lett.*, 47(26):1927–1930.

Sarychev, Andrey K., Shvets, Gennady, and Shalaev, Vladimir M. (2006). Magnetic plasmon resonance. *Phys. Rev. E*, 73:036609.

Sauer, G., Brehm, G., Schneider, S., Graener, H., Seifert, G., Nielsch, K., Choi, J., Göring, P., Gösele, U., Miclea, P., and Wehrspohn, R. B. (2005). In situ surface-enhanced Raman spectroscopy of monodisperse silver nanowire arrays. *J. Appl. Phys.*, 97:024308.

Sauer, G., Brehm, G., Schneider, S., Graener, H., Seifert, G., Nielsch, K., Choi, J., Göring, P., Gösele, U., Miclea, P., and Wehrspohn, R. B. (2006). Surface-enhanced Raman spectroscopy employing monodisperse nickel nanowire arrays. *Appl. Phys. Lett.*, 88:023106.

Saxler, J., Rivas, J. Gomez, Janke, C., Pellemans, H. P. M., Bolivar, P. Haring, and Kurz, H. (2004). Time-domain measurements of surface plasmon polaritons in the terahertz frequency range. *Phys. Rev. B*, 69:155427.

Schouten, H. F., Kuzmin, N., Dubois, G., Visser, T. D., Gbur, G., Alkemade, P. F. A., Blok, H., 't Hooft, G. W., Lenstra, D., and Eliel, E. R. (2005). Plasmon-assisted two-slit transmission: Young's experiment revisited. *Phys. Rev. Lett.*, 94:053901.

Shalaev, Vladimir M. (2000). *Nonlinear optics of random media*. Springer, Heidelberg, Germany, first edition.

Shalaev, Vladimir M., Cai, Wenshan, Chettiar, Uday K., Yuan, Hsiao-Kuan, Sarychev, Andrey K., Drachev, Vladimir P., and Kildishev, Alexander V. (2005). Negative index of refraction in optical metamaterials. *Opt. Lett.*, 30(24):3356–3358.

Shalaev, Vladimir M., Poliokov, E. Y., and Markel, V. A. (1996). Small-particle composites. II. Nonlinear optical properties. *Phys. Rev. B*, 53(5):2437–2449.

Shelby, R., Smith, D. R., and Schultz, S. (2001). Experimental verification of a negative index of refraction. *Science*, 292:77–79.

Shi, Xiaolei, Hesselink, Lambertus, and Thornton, Robert L. (2003). Ultrahigh light transmission through a C-shaped nanoaperture. *Opt. Lett.*, 28(15):1320–1322.

Shin, Hocheol, Catrysse, Peter B., and Fan, Shanhui (2005). Effect of the plasmonic dispersion relation on the transmission properties of subwavelength cylindrical holes. *Phys. Rev. B*, 72:085436.

Shou, Xiang, Agrawal, Amit, and Nahata, Ajay (2005). Role of metal film thickness on the enhanced transmission properties of a periodic array of subwavelength apertures. *Opt. Express*, 13(24):9834–9840.

Shubin, V. A., Kim, W., Safonov, V. P., Sarchev, A. K., Armstrong, R. L., and Shalaev, Vladimir M. (1999). Surface-Plasmon-Enhanced Radiation Effects in Confined Photonic Systems. *J. Lightwave Tech.*, 17(11):2183–2190.

Simon, H. J., Mitchell, D. E., and Watson, J. G. (1974). Optical second-harmonic generation with surface plasmons in silver films. *Phys. Rev. Lett.*, 33(26):1531–1534.

Sipe, J. E., So, V. C. Y., Fukui, M., and Stegeman, G. I. (1980). Analysis of second-harmonic generation at metal surfaces. *Phys. Rev. B*, 21(10):4389–4402.

Slavik, Radan, Homola, Jiri, and Ctyroky, Jiri (1999). Single-mode optical fiber surface plasmon resonance sensor. *Sensors and Actuators B*, 74:74–79.

Smith, D. R., Pendry, J. B., and Wiltshire, M. C. K. (2004). Metamaterials and negative refractive index. *Science*, 305:788–792.

Smolyaninov, Igor I., Hung, Yu-Ju, and Davis, Christopher C. (2005). Surface plasmon dielectric waveguides. *Appl. Phys. Lett.*, 87:241106.

Sommerfeld, A. (1899). Über die Fortpflanzung electrodynamischer Wellen längs eines Drahtes. *Ann. Phys. und Chemie*, 67:233–290.

Sönnichsen, C., Franzl, T., Wilk, T., von Plessen, G., and Feldmann, J. (2002a). Plasmon resonances in large noble-metal clusters. *New Journal of Physics*, 4:93.1–93.8.

Sönnichsen, C., Franzl, T., Wilk, T., von Plessen, G., Feldmann, J., Wilson, O., and Mulvaney, P. (2002b). Drastic reduction of plasmon damping in gold nanorods. *Phys. Rev. Lett.*, 88(7):077402.

Sönnichsen, C., Geier, S., Hecker, N. E., von Plessen, G., Feldmann, J., Ditlbacher, H., Lamprecht, B., Krenn, J. R., Aussenegg, F. R., Chan, V. Z., Spatz, J. P., and Moller, M. (2000). Spectroscopy of single metal nanoparticles using total internal reflection microscopy. *Appl. Phys. Lett.*, 77:2949–2951.

Sönnichsen, Carsten and Alivisatos, A. Paul (2005). Gold nanorods as novel nonbleaching plasmon-based orientation sensors for polarized single-particle spectroscopy. *Nano Letters*, 5(2):301–304.

Spillane, S. M., Kippenberg, T. J., and Vahala, K. J. (2002). Ultralow-threshold Raman laser using spherical dielectric microcavity. *Nature*, 415:621–623.

Srituravanich, Werayut, Fang, Nicholas, Sun, Cheng, Luo, Qi, and Zhang, Xiang (2004). Plasmonic nanolithography. *Nano Lett.*, 4(6):1085–1088.

Stegeman, G. I., Wallis, R. F., and Maradudin, A. A. (1983). Excitation of surface polaritons by end-fire coupling. *Opt. Lett.*, 8(7):386–388.

Stern, E. A. and Ferrell, R. A. (1960). Surface plasma oscillations of a degenerate electron gas. *Phys. Rev.*, 120(1):130–136.

Stockman, Martin I. (2004). Nanofocusing of optical energy in tapered plasmonic waveguides. *Phys. Rev. Lett.*, 93(13):137404.

Su, K.0H., Wei, Q.-H., Zhang, X., Mock, J. J., Smith, D. R., and Schultz, S. (2003). Interparticle coupling effects on plasmon resonances of nanogold particles. *Nano Letters*, 3(8):1087–1090.

Sundaramurthy, Arvind, Crozier, K. B., Kino, G. S., Fromm, D. P., Schuck, P. J., and Moerner, W. E. (2005). Field enhancement and gap-dependent resonance in a system of two opposing tip-to-tip Au nanotriangles. *Phys. Rev. B*, 72:165409.

Takahara, J., Yamagishi, S., Taki, H., Morimoto, A., and Kobayashi, T (1997). Guiding of a one-dimensional optical beam with nanometer diameter. 22(7):475–477.

Talley, Chad E., Jackson, Joseph B., Oubre, Chris, Grady, Nathaniel K., Hollars, Christopher W., Lane, Stephen M., Huser, Thomas R., Nordlander, Peter, and Halas, Naomi J. (2005). Surface-enhanced Raman scattering from individual Au nanoparticles and nanoparticle dimer substrates. *Nano Letters*, 5(8):1569–1574.

Tam, Felicia, Moran, Cristin, and Hallas, Naomi (2004). Geometrical parameters controlling sensitivity of nanoshell plasmon resonances to changes in dielectric environment. *J. Phys. Chem. B*, 108:17290–17294.

Tan, W.-C., Preist, T. W., and Sambles, R. J. (2000). Resonant tunneling of light through thin metal films via strongly localized surface plasmons. *Phys. Rev. B*, 62(16):11134–11138.

Tanaka, Kazuo and Tanaka, Masahiro (2003). Simulations of nanometric optical circuits based on surface plasmon polariton gap waveguide. *Appl. Phys. Lett.*, 82(8):1158–1160.

Teperik, T. V. and Popov, V. V. (2004). Radiative decay of plasmons in a metallic nanoshell. *Phys. Rev. B*, 69:155402.

Thio, Tineke, Pellerin, K. M., Linke, R. A., Lezec, H. J., and Ebbesen, T. W. (2001). Enhanced light transmission through a single subwavelength aperture. *Opt. Lett.*, 26(24):1972–1974.

Tian, Zhong-Qun and Ren, Bin (2004). Adsorption and reaction of electrochemical interfaces as probed by surface-enhanced Raman spectroscopy. *Annu. Rev. Phys. Chem.*, 55:197–229.

Tsai, Woo-Hu, Tsao, Yu-Chia, Lin, Hong-Yu, and Sheu, Bor-Chiou (2005). Cross-point analysis for a multimode fiber sensor based on surface plasmon resonance. *Opt. Lett.*, 30(17):2209–2211.

van der Molen, K. L., Segerink, F. B., van Hulst, N. F., and Kuipers, L. (2004). Influence of hole size on the extraordinary transmission through subwavelength hole arrays. *Appl. Phys. Lett.*, 85(19):4316–4318.

van Exter, Martin and Grischkowsky, Daniel R. (1990). Characterization of an optoelectronic terahertz beam system. *IEEE Transactions on Microwave Theory and Techniques*, 38(11):1684–1691.

Veronis, Georgios and Fan, Shanhui (2005). Guided subwavelength plasmonic mode supported by a slot in a thin metal film. *Opt. Lett.*, 30(24):3359–3361.

Vial, Alexandre, Grimault, Anne-Sophie, Macías, Demetrio, Barchiesi, Dominique, and de la Chapelle, Marc Lamy (2005). Improved analytical fit of gold dispersion: Application to the modeling of extinction spectra with a finite-difference time-domain method . *Phys. Rev. B*, 71:085416.

Vincent, R. and Silcox, J. (1973). Dispersion of radiative surface plasmons in aluminum films by electron scattering. *Phys. Rev. Lett.*, 31(25):1487–1490.

Vuckovic, J., Loncar, M., and Scherer, A. (2000). Surface plasmon enhanced light-emitting diode. *IEEE J. Quan. Elec.*, 36(10):1131–1144.

Wachter, M., Nagel, M., and Kurz, H. (2005). Frequency-dependent characterization of THz Sommerfeld wave propagation on single-wires. *Opt. Express*, 13(26):10815–10822.

Wait, James R. (1998). The ancient and modern history of EM ground-wave propagation. *IEEE Antennas and Propagation Magazine*, 40(5):7–24.

Wang, Kanglin and Mittleman, Daniel M. (2005). Metal wires for terahertz wave guiding. *Nature*, 432:376–379.

Wang, Qian-jin, Li, Jia-qi, Huang, Cheng-ping, Zhang, Chao, and Zhu, Yong-Yuan (2005). Enhanced optical transmission through metal films with rotation-symmetrical hole arrays. *Appl. Phys. Lett.*, 87:091105.

Watanabe, Hiroyuki, Ishida, Yasuhito, Hayazawa, Norihiko, Inouye, Yasishi, and Kawata, Satoshi (2004). Tip-enhanced near-field Raman analysis of tip-pressurized adenine molecule. *Phys. Rev. B*, 69:155418.

Watts, Richard A. and Sambles, J. Roy (1997). Polarization conversion from blazed diffraction gratings. *J. Mod. Opt.*, 44(6):1231–1242.

Webb, K. J. and Li, J. (2006). Analysis of transmission through small apertures in conducting films. *Phys. Rev. B*, 73:033401.

Weber, W. H. and Ford, G. W. (2004). Propagation of optical excitations by dipolar interactions in metal nanoparticle chains. *Phys. Rev. B*, 70:125429.

Wedge, S., Hooper, I. R., Sage, I., and Barnes, W. L. (2004). Light emission through a corrugated metal film: The role of cross-coupled surface plasmon polaritons. *Phys. Rev. B*, 69:245418.

Weeber, J.-C., Krenn, J. R., Dereux, A., Lamprecht, B., Lacroute, Y., and Goudonnet, J. P. (2001). Near-field observation of surface plasmon polariton propagation on thin metal stripes. *Phys. Rev. B*, 64:045411.

Weeber, Jean-Claude, Lacroute, Yvon, and Dereux, Alain (2003). Optical near-field distributions of surface plasmon waveguide modes. *Phys. Rev. B*, 68:115401.

Weeber, Jean-Claude, Lacroute, Yvon, Dereux, Alain, Devaux, Eloise, Ebbesen, Thomas, Girard, Christian, Gonzalez, Maria Ujue, and Baudrion, Anne-Laure (2004). Near-field characterization of Bragg mirrors engraved in surface plasmon waveguides. *Phys. Rev. B*, 70:235406.

Weitz, D. A., Garoff, S., Gersten, J. I., and Nitzan, A. (1983). The enhancement of Raman scattering, resonance Raman scattering, and fluorescence from molecules adsorbed on a rough silver surface. *J. Chem. Phys.*, 78(9):5324–5338.

Westcott, S. L., Jackson, J. B., Radloff, C., and Halas, N. J. (2002). Relative contributions to the plasmon line shape of metallic nanoparticles. *Phys. Rev. B*, 66:155431.

Wilcoxon, J. P. and Martin, J. E. (1998). Photoluminescence from nanosize gold clusters. *J. Chem. Phys.*, 108(21):9137–9143.

Winter, G. and Barnes, W. L. (2006). Emission of light through thin silver films via near-field coupling to surface plasmon polaritons. *Appl. Phys. Lett.*, 88:051109.

Wokaun, A., Gordon, J. P., and Liao, P. F. (1982). Radiation damping in surface-enhanced Raman scattering. *Phys. Rev. Lett.*, 48(14):957–960.

Wood, R. W. (1902). On a remarkable case of uneven distribution of light in a diffraction grating spectrum. *Proc. Phys. Soc. London*, 18:269–275.

Wurtz, Gregory A., Im, Jin Seo, Gray, Stephen K., and Wiederrecht, Gary P. (2003). Optical scattering from isolated metal nanoparticles and arrays. *J. Phys. Chem. B*, 107(51):14191–14198.

Xu, Hongxing (2004). Theoretical study of coated spherical metallic nanoparticles for single-molecule surface-enhanced spectroscopy. *Appl. Phys. Lett.*, 85(24):5980–5982.

Xu, Hongxing, Aizpurua, Javier, Kaell, Mikael, and Apell, Peter (2000). Electromagnetic contributions to single-molecule sensitivity in surface-enhanced Raman scattering. *Phys. Rev. E*, 62(3):4318–4324.

Xu, Hongxing, Wang, Xue-Hua, and Käll, Mikael (2002). Surface-plasmon-enhanced optical forces in silver nanoaggregates. *Phys. Rev. Lett.*, 89(24):246802.

Xu, Hongxing, Wang, Xue-Hua, Persson, Martin P., Xu, H. W., and Käll, Mikael (2004). Unified treatment of fluorescence and Raman scattering processes near metal surfaces. *Phys. Rev. Lett.*, 93:243002.

Yamamoto, N., Araya, K., and de Abajo, F. J. García (2001). Photon emission from silver particles induced by a high-energy electron beam. *Phys. Rev. B*, 64:205419.

Yariv, Amnon (1997). *Optical Electronics in Modern Communications*. Oxford Univeristy Press, Oxford, UK, fifth edition edition.

Yen, T. J., Padilla, W. J., Fang, N., Vier, D. C., Smith, D. R., Pendry, J. B., Basov, D. N., and Zhang, X. (2004). Terahertz magnetic response from artificial materials. *Science*, 303:1494–1496.

Yin, L., Vlasko-Vlasov, V. K., Rydh, A., Pearson, J., Welp, U., Chang, S.-H., Gray, S. K., Schatz, G. C., Brown, D. B., and Kimall, C. W. (2004). Surface plasmons at single nanoholes in Au films. *Appl. Phys. Lett.*, 85(3):467–469.

Yin, Leilei, Vlasko-Vlasov, Vitali K., Pearson, John, Hiller, Jon M., Hua, Jiong, Welp, Ulrich, Brown, Dennis E., and Kimball, Clyde W. (2005). Subwavelength focusing and guiding of surface plasmons. *Nano Letters*, 5(7):1399–1402.

Zayats, A. V. and Smolyaninov, I. I. (2006). High-optical throughput individual nanoscale aperture in a multilayered metallic film. *Opt. Lett.*, 31(3):398–400.

Zenneck, J. (1907). Über die Fortpflanzung ebener elektromagnetischer Wellen längs einer ebenen Leiterfläche und ihre Beziehung zur drahtlosen Telegraphie. *Ann. d. Phys.*, 23:846–866.

Zhang, Y., Gu, C., Schwartzberg, A. M., and Zhang, J. Z. (2005). Surface-enhanced Raman scattering sensor based on D-shaped fiber. *Appl. Phys. Lett.*, 87:123105.

Zhou, J., Koschny, Th., Kafesaki, M., Economou, E. N., Pendry, J. B., and Soukoulis, C. M. (2005). Saturation of the magnetic response of split-ring resonators at optical frequencies. *Phys. Rev. Lett.*, 95:223902.

Zia, Rashid, Chandran, Anu, and Brongersma, Mark L. (2005a). Dielectric waveguide model for guided surface polaritons. *Opt. Lett.*, 30(12):1473–1475.

Zia, Rashid, Selker, Mark D., and Brongersma, Mark L. (2005b). Leaky and bound modes of surface plasmon waveguides. *Phys. Rev. B*, 71:165431.

Zia, Rashid, Selker, Mark D., Catrysse, Peter B., and Brongersma, Mark L. (2005c). Geometries and materials for subwavelength surface plasmon modes. *J. Opt. Soc. Am. A*, 21(12):2442.

Index